Gorillas & Chimpanzees

R. L. Garner

Alpha Editions

This edition published in 2022

ISBN : 9789356153189

Design and Setting By
Alpha Editions
www.alphaedis.com
Email - info@alphaedis.com

Contents

PREFACE

The present work is the natural product of some years devoted to a study of the speech and habits of monkeys. It has led up to the special study of the great apes. The matter contained herein is chiefly a record of the facts tabulated during recent years in that field of research.

The aim in view is to convey to the casual reader a more correct idea than now prevails concerning the physical, mental, and social habits of these apes.

The favourable conditions under which the writer has been placed, in the study of these animals in the freedom of their native jungle, have not hitherto been enjoyed by any other student of Nature.

A careful aim to avoid all technical terms and scientific phraseology has been adhered to, and the subject treated in a simple style. Tedious details are relieved by an ample supply of anecdotes taken from the writer's own observations, and most of them are the acts of his own pets or of apes in a wild state. The author has refrained from rash deductions and abstruse theories, but has sought to place the animals here treated in their true light, believing that to dignify the apes is not to degrade man, but to exalt him even more.

It is hoped that a more perfect knowledge of these animals may bring man into closer fellowship and deeper sympathy with Nature, and cause him to realise that all creatures think and feel in some degree, however small.

THE AUTHOR.

CHAPTER I
MAN AND APE COMPARED

Monkeys have always been a subject of idle interest to old and young; but they have usually served to amuse the masses more than to instruct them, until within recent years.

Now that science has brought them within the field of careful research, and made them an object of serious study, it has invested them with a certain dignity in the esteem of mankind, and imparted to them a new aspect among animals.

There is no other creature that so charms and fascinates the beholder as do these little effigies of the human race. The simple and the wise are alike impressed with their human look and manner; children and patriarchs with equal delight watch them with surprise; but now that the search-light of science is being thrown into every nook and crevice of nature, human interest in them is multiplied many fold, while the savants of all civilised lands are struggling with the problem of their possible relationship to man.

Pursuant to the desire of learning as much as possible about their natural habits, faculties, and resources, they are being studied from every available point of view, and every characteristic compared in detail to the corresponding one in man. Hence, in order to appreciate more fully the value of the lessons to be drawn from the contents of this volume, we must know the relative planes in the scale of nature that man and monkeys occupy, wherefore we shall begin our task by comparing them in a general way; but as the scope of this work is restricted mainly to the great apes, the comparison will likewise be confined to that subject, except in so far as to define the relations of man and ape to monkeys.

Since monkeys differ among themselves so widely, it is evident that all of them cannot in the same degree resemble man. And as the degree of interest in them as a subject of comparative study is approximately measured by the degree of their likeness to man, it is apparent that all cannot be regarded as of equal interest. But since each forms an integral part of the scale of nature, they are of equal importance in tracing out the continuity of the order to which they belong.

The vast family of simians has perhaps the widest range of types of any single family of mammals. Beginning with the great apes, which so closely resemble man in size, form and structure, they descend by degrees along the scale till they end in the little marmosets, which are almost on the level of rodents. But the descent is so gradual that it is difficult to draw a sharp line of demarcation at any point between the two extremes. There is, however, now an effort being made to separate this family into smaller groups, but the lines between them must be dim and wavering, and the literature of the past has a tendency to retard the effort.

We shall not digress from the trend of our subject, however, at this time, to discuss the problems with which zoology may have to contend in the future, but will accept the current system and proceed.

All the varied types that belong to the simian family are, in the common order of speech, known as *monkeys*, but the term thus used is so broad in its meaning as to include all the forms of that vast group, wherefore it is vague and obscure, for some of these resemble man more than they resemble each other. The name should only be applied to those having tails and short faces, but there is a small group, which have no tails at all, that are properly known as *apes*. While they are all simians, they are not all monkeys. It is with this small group, without tails, that we propose chiefly to deal. We select them because of their likeness to man, and having noted the similitude, the result may be compared with other types of the same order. There are only four of these apes, but as a whole they resemble man in so many essential details that they are called "anthropoid," or "man-like apes." They differ from each other in certain respects, almost as much as any one of them differs from man. The four apes alluded to, are the chimpanzee, the gorilla, the orang and the gibbon.

As the skeleton is the framework of the physical structure, it will serve as the basis upon which to build up the comparison, and as the chimpanzee is the nearest approach to man, we select him as the highest type of the simian, and use him as the standard.

The skeleton of the chimpanzee may be said to be exactly the same as that of man, but the assertion must be qualified by a few facts which are of minor importance, but since they are facts we cannot ignore them.

The general plan, purpose and principle are the same in each. There is no part of the one that is not duplicated in the other, and there is no function discharged by any part of the one that is not discharged by the like part of the other. The chief point in which they differ is in the structure of one bone.

Near the base of the spinal column is a certain bone called the *sacrum*. It is a constituent part of the column, but in its singular form and structure somewhat differs from the corresponding bone in man. The general outline of this bone in the plane of the hips is that of an isosceles triangle. It fits in between the two large bones that spread out towards the hips, and articulate with the thighbones.

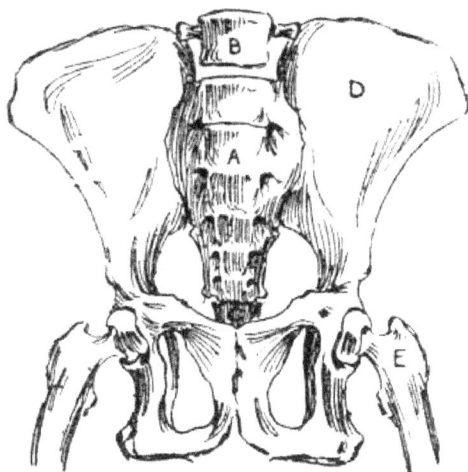

PELVIS OF CHIMPANZEE

A Sacrum.
B Fourth lumbar vertebra.
C Coccyx.
D Ilium or hip-bone.
E Femur or thigh-bone.

About half-way from the centre to the edge, along each side, is a row of four round holes. Across the surface of the bone is a dim transverse line between each pair of holes, from which it appears that five smaller sections of the column have anchylosed or grown into each other to form the *sacrum*, and the holes coincide with the open spaces between the lateral processes of the other bones of the column above.

In the chimpanzee, this bone has the same general form as in man, but instead of four holes in each row it has five, connected by transverse lines in the same way, indicating that six of the segments are united instead of five; but to compensate for this the ape has one vertebra less in the section of the column just above it, in that portion called the *lumbar*. In it man has five, while the ape has but four. But counting the whole number of bones in the spinal column, and regarding each segment of the *sacrum* as a distinct bone, which to all intents it is, the sum of the bones in each column is exactly the same.

Although this appears to be a fixed and constant character, it cannot be esteemed as a matter of great importance, since the same thing has been known to occur in the human skeleton, and the reverse has been known in some specimens of the apes, but has never been observed in the chimpanzee.

In this respect he appears to be more constant than man so far as we know at present.

As the greatest strains of the spinal column are laid upon that part in which the *sacrum* is located, there is a tendency for these segments to unite in order to meet the demand, and since there is the least flexure in that part, the cartilages that lie between them ossify and become rigid. The erect posture of man allows more room in the loins for the fifth vertebra to move, and thus it is prevented from uniting with the segment below it, which is held firmly in place by the two large bones mentioned, while the crouching habit of the ape presses that vertebra firmly against the other, confining it between the two large bones and thus reducing its movement, wherefore the same result follows as with the other sections below.

Another bone that may be said to differ in structure is that known as the *sternum* or breastbone; it is the thin, soft bone to which the ribs are joined in the front of the body. In the young of both man and ape it is a mere cartilage which slowly ossifies from the top downward. The process appears to begin at different centres, the largest nucleus being at the top. There appear to be five of these centres. The bone never becomes quite hard in either man or ape, but always remains somewhat porous, and even in advanced age the outline of the lower part is not defined by a smooth, sharp line, but is irregular in contour and merges or blends into the cartilages that hold the ribs in place.

In man, this bone in maturity is usually found in two segments, while in the ape it varies. In some specimens it is the same as in man, while in others it is found to be in four or five segments. But the *sternum* in each is always regarded as one bone, and is developed from one continuous cartilage. The separate parts are never considered distinct bones. The reason that it is found in separate sections in the ape is doubtless due to the stooping habit of the animal, by which the bone is constantly flexed and alternately straightened. In man this bone varies to a great extent.

With these trifling exceptions in point of structures alone, the skeletons of man and ape may be truly said to be exact counterparts of each other, having the same number of bones, of the same general type arranged in the same order and articulated in the same manner. The corresponding bone in each is the same in design and purpose. The frame of the ape is much more massive in its proportions than that of man, but while this is true of some kinds of ape the reverse is true of others. The average height of the adult chimpanzee is about 63 inches.

In man the *sacrum* is more curved in the plane of the hips than it is in the ape, while the bones of the digits in man are straighter. The arms of man are shorter than the legs, while in the ape these features are reversed.

In the cranial types, it is readily seen that the skull of man is nearly round and the face is vertical, while the skull of the ape is elongated and the face receding. These facts deserve more notice than the mere mention of their being so.

In the whole scheme of nature certain laws obtain in the projection of skulls. The angle between the plane of the face and the spinal axis is co-ordinate to the angle between the spinal axis and the perpendicular.

To be more exact, the spine of a snake is in a horizontal line, and the face occupies a plane of the same kind. At the other end of the scale is man, whose spine is in a vertical line, and his face occupies a like plane. Between these two extremes are types which tend in various degrees, from the lower to the higher form, and just in proportion as the spinal axis approaches a vertical line from one side, the plane of the face approaches it from the other.

In accord with this fact it will be observed that the foramen or hole in the base of the skull through which the spinal cord passes is adjusted closer and closer to the centre of the base of the skull as the spine becomes erect. In man, whose spinal column is erect, the hole is in the centre of the base; in the reptile, whose spine is horizontal, the hole is at the extreme end of the base. In the ape the spinal axis is at an angle with the vertical line, and the plane of the face conforms to a similar one. In keeping with this law it will be seen in all animals that just in the same degree as the angles widen, the foramen is removed from the centre of the base towards the occiput.

It may be noted here, however, that the facial angle is never exactly the same as the spinal angle. The facial plane of the reptile is not quite horizontal, nor that of man quite vertical, but the ratios of angularity are constant. Even the habit of rearing modifies to some extent this character, but it is only the normal pose of the animal that determines the exact limit of it.

In keeping with these facts it will be observed that as the angle between the chin and the spine widens, the lower jaws project, and the chin recedes or flattens, and in a like degree the voice is modified. The chin of man forms a right angle, but in the reptile it is quite lost. In the former the vocal powers are superior to that of all other animals, but as we descend the scale they are reduced in scope and degraded in quality, until in the lowest reptiles they become a mere hiss or squeak.

By a careful study of the voices together with the skulls of animals, it is found that the gnathic index can be relied upon as a vocal index. The ape has the smallest angle between the spinal axis and the facial plane, and has the greatest vocal range and purest voice of any other animal below man. Among the apes the gibbon has the smallest angle, and he also has the best vocal qualities of any other ape.

The contour of the skull in all parts conforms to the angle of its projection from the spinal axis. It is depressed and elongated in proportion as the angle increases: the brain cavity is narrowed in a like proportion to its length, and the brain, of course, is modified in the same manner.

The brain of the ape resembles that organ in man as closely as his skeleton resembles man's. It has the same lobes, convolutions, and centres. The texture is slightly coarser. The small details are less intricate and their lines somewhat less distinct. But these also differ to a certain extent in different men. In man and apes the same nerves are present and connect the same organs of sensation, volition and motion. In all essential points they are one.

These leading facts are deemed sufficient to show the physical likeness of apes to man, and we shall refrain from the minute details that would only be of interest to the specialist. The purpose is to acquaint the general reader with the leading facts.

Regarding man purely in the light of an animal, it is evident that he is, physically, very closely allied to the chimpanzee, and that both are integral parts of one great scheme of life, designed by the same author, fashioned after the same model, projected upon the same plan, and amenable to the same system of vital economy. Viewing him in the light of his physical nature, so far it is found that he does not materially differ from other animals in the structure of his skeleton and certain concomitants.

In the vital organs of the two there is perhaps still greater unity of structure, and equal unity of function in all essential details. The difference of structure is only to the extent of making the organ conform to the general plan of the animal, and the difference of function is only one of degree. Since the same characters vary quite as much among men without changing their identity as such, it cannot be sufficient ground to widen the hiatus between man and ape; in fact, the physical likeness of the two grows stronger as the comparison is extended into more minute and scrutinising details. To the casual observer the general resemblance is apparent, but to the student the unity becomes evident.

In addition to the facts we have cited, the ape has the same habits of rest and sleep; lives on the same kind of diet, which is eaten and assimilated in the same manner as with man; is subject to many of the same diseases which attack the same organs, and affect them in the same way as with man; he suffers like pains and dies in the same manner as man under like conditions.

The scope of this book is intended only to embrace the chimpanzee and gorilla, but the comparison which we have shown applies in the name to all four of the anthropoid apes, but must be qualified in a few instances to make it apply to the others. These apes differ among themselves in certain respects in form and habits, and we will omit a detailed comparison of the monkeys as not being relevant to the subject in hand; but it will not be out of place to mention in a general way the chief point in which they differ from men and apes.

There is no fixed type that will represent all kinds of monkeys.

Within the limits of their own family they present a great variety of types, but the one marked difference between them as a unit, and the ape as another, is, that the spinal column of the monkey is always extended into a tail, the first vertebra of which is joined to the base of the *sacrum*, while the ape has no tail, but the spinal column terminates with a small pointed bone called the coccyx, exactly the same as in man. The number of bones and the number of ribs in monkeys differ from those in the ape or in man, and also vary among different types of monkey.

There are many little shades and grades of difference all along the line, but the unity of design throughout the whole range of simian life is such as to show a continuity of plan and purpose in all essential details of the animal economy. With man and ape the physical structures are one, so far as they pertain to autonomy: their habits are one, so far as they pertain to the means of life; their faculties are one, so far as they pertain to the animal polity, yet they may not be of a common stock.

The public mind does not seem to have grasped the correct idea of evolution, and prejudice has blinded, to some extent, the judgment. The common opinion that man has descended from or is related by consanguinity to a monkey is silly and absurd. Science has never taught such folly, nor advanced any theory from which such a conclusion could be justly deduced. It would be a waste of time for me to offer to explain the doctrine of evolution to any one who does not already understand it from the literature of others on this subject. If he still nurse the idol of the identity of man and monkey, he must be too obtuse or too perverse to be reclaimed. But no one will deny the physical resemblance between man and the great apes, and it is this resemblance we seek to show rather than trace any relationship based upon theories. It is not a matter that concerns the purpose of this work, and we shall here dismiss the subject by saying, that things may be equivalent and yet not identical.

CHAPTER II
CAGED IN AN AFRICAN JUNGLE

It may be of interest to the reader to know the manner in which I have pursued the study of monkeys in a state of nature, and the means employed to that end. I shall therefore give a brief outline of my life in a cage in the heart of an African jungle in order to watch those denizens of the forest, when free from all restraint.

After devoting much time for several years to the study of the speech and habits of monkeys in captivity, I formulated a plan of going into their native haunts, to study them in a state of freedom.

In the course of my labours up to that time, I had found out that monkeys of the highest physical type had also a higher type of speech than those of inferior kinds. In accord with this fact, it was logical to infer that the anthropoid apes, being next to man in the scale of nature, must have the faculty of speech developed in a corresponding degree.

As the chief object of my studies was to learn the language of monkeys, the great apes appeared to be the best subjects for that purpose, so I turned my attention to them.

The gorilla was said to be the most like man, and the chimpanzee next. There were none of the former in captivity, and but few of the latter, and they were kept under conditions that forbade all efforts to do anything in that line.

As the gorilla and chimpanzee could both be found in the same section of tropical Africa, I selected that as the field of operation, and began to prepare for a journey there to carry out the task I had assumed.

The part selected was along the equator, and south of it, about two degrees. The locality is infested with fevers, insects, serpents and wild beasts of divers kinds. To ignore such dangers would be folly, but there was no way to see these apes in their freedom, except to go and live among them.

To lessen, in a degree, the dangers incurred by such an adventure, I devised a cage of steel wire, woven into a lattice with a mesh one inch and a half wide. This was made in twenty-four panels, three feet three inches square, set in a frame of narrow iron strips. Each side of the panels was provided with half-hinges, so arranged as to fit any side of every other panel. These could be quickly bolted together with small iron rods, and, when so bolted, formed a cage of cubical shape, six feet six inches square.

Any one or more of the panels could be swung open as a door, and the whole structure was painted a dingy green, so that when erected in the forest it was almost invisible among the foliage.

While it was not strong enough to withstand a prolonged siege, it afforded a certain immunity from being surprised by the fierce and stealthy beasts of the jungle, and would allow the occupant time to kill an assailant before the wires would yield to an attack from anything except an elephant. Of course it was no protection against them, but as they rarely ever attack a man unless provoked to it, there was little danger from that source; besides, there were not many of those huge brutes in the immediate part in which my strange domicile was set up.

Through this open fabric one could see without obstruction on all sides, and yet feel a certain sense of safety from being devoured by leopards or panthers.

Over this frail fortress was a roof of bamboo leaves, and it was provided with curtains of canvas to be hung up in case of rain. The floor was of thin boards, steeped in tar, and the structure was set up about two feet from the ground, on nine small posts.

WAITING AND WATCHING IN THE CAGE

It was furnished with a bed, made of heavy canvas supported by two poles of bamboo, attached to the edge of it. One of these poles was lashed fast to the side of the cage, and the other was suspended at night by strong wire hooks, hung on the top of it. During the day, the bed was rolled up on one of the poles, so that it was out of the way. I had a light camp chair, which folded up, and a table was improvised by a broad, short board hung on wires. This could be set up by the wall of the cage at night, out of the way. To this meagre outfit was added a small kerosene stove, and a swinging shelf.

A few tin cases contained my wearing apparel, blanket, pillow, photograph camera and supplies, medicines, and an ample store of canned meats, crackers, &c. A magazine rifle, revolver, ammunition, and a few useful tools, such as a hammer, saw, pliers, files, and a heavy bush-knife, completed my stock, except some tin platters, cups and spoons. These served in cooking, and also for the table, instead of dishes.

With this equipment I sailed from New York on the 9th of July 1892, *viâ* England, to the port of Gaboon, the site of the colonial government of the French Congo. This place is within a few miles of the equator, and near the borders of the country in which the gorilla lives. I arrived there on the 18th of October of the same year, and after a delay of a few weeks I set out to find the object of my search.

Leaving this place, I went up the Ogowe River about two hundred miles, and through the lake region on the south side of it. After some weeks of travel and inquiry, I arrived at the lake of Ferran Vaz, in the territory of the Nkami tribe. The lake is about thirty miles long, by eight or ten wide, and interspersed with a few islands of large size, covered with a dense growth of tropical vegetation. The country around the lake is mostly low and marshy, traversed by creeks, lagoons and rivers. Most of the land is covered by a deep and dreary jungle, with a few sandy plains at intervals.

In the depths of this gloomy forest, reeking with the effluvia of decaying plants, and teeming with insect life, the gorilla dwells in safety and seclusion. In the same forest the chimpanzee makes his abode, but is less timid and retiring.

On the south side of this lake, not quite two degrees below the equator, and within some twenty miles of the ocean, I selected a place in the heart of the primeval forest, erected my little fortress, and gave it the name of *Fort Gorilla.*

In the latter part of April 1893, I took up my abode in this desolate spot, and began my long and solitary vigil.

My sole companion was a young chimpanzee, that I named Moses, and, from time to time, a native boy, as a servant.

Seated in this cage, in the silence of the great forest, I have seen the gorilla in all his majesty, strolling at leisure through his sultry domain, in quest of food. I have seen the chimpanzee under like conditions, and the happy, chattering monkey in the freedom of his jungle home.

In this novel hermitage I remained for the greater part of the time for one hundred and twelve days and nights in succession, watching these animals in perfect freedom following the pursuits of their daily life.

With such an experience, I will not be charged with vanity in saying that I have seen more of those animals in a state of nature than any white man ever saw, and under conditions more favourable for a careful study of their manners and habits, than could otherwise be possible. Hence, what I have to say concerning them is the result of an experience which no other man can claim.

I do not mean to ignore or impugn what others have said on this subject, but the sum of my labours in this field leads me to doubt much that has been said and accepted as true. I regret that it devolves upon me to controvert many stories told about these great apes, but finding no germ of truth in some of them, I cannot evade the duty of denying them. I regret it all the more, because many of them have been woven into the fabric of natural history, and marked with the seal of scientific approval; but time will sustain me in the denial.

I am aware that bigots of certain schools will challenge me for pointing out their mistakes, and some will assume to know more about these apes than a fish knows about swimming; but truth defies all theory.

Each kind of ape will be treated in the chapter devoted to it, but only those with which I have dealt in person will be discussed at length. Others will be noticed, in order to sustain the continuity of the subject and show the relative planes of those under consideration. But before proceeding with the monkeys, I shall pause to relate some of the incidents of my hermitage.

CHAPTER III
DAILY LIFE AND SCENES IN THE JUNGLE

I am so frequently asked about the details of my daily life in the cage, how the time was occupied and what I saw besides the apes, that I deem it of interest to relate a few of the events of my sojourn in this wild spot.

In order to convey an idea of it, I shall relate the incidents of a single day and night; but of course the routine varied in some degree from day to day.

At six o'clock, as the sun first peeps into the forest, it finds me with a tin cup of coffee, just made on the little kerosene stove. It is black and dreggy, but with a little sugar it is not bad. With a few dry crackers I break my fast of twelve hours, and am ready for the task before me.

STARTING FOR A STROLL

In the meantime the boy rolls up my bed and his mat. By this time Moses has helped himself to a banana or two. Then I take my rifle, he climbs up on my shoulder, and we go for a short walk in the bush, while the boy sweeps out the cage and puts everything in order for the day. When we return, the boy, armed with a native spear, or a huge knife, takes the big jug, and goes to a spring, about three hundred yards away, for a supply of water.

Then Moses is allowed to climb about in the bushes and amuse himself; the boy sits down, or goes to his village a mile away, while I watch for gorillas. Silence is the order of the day, and here I sit, sometimes for hours alone, almost as quiet as a tomb.

Presently a rustle of the leaves is heard, and a porcupine comes waddling into view. He is poking his nose about, in search of food, but has not discovered my presence. He comes closer, until the scent or sight of me startles him, and away he goes. By-and-by a civet cat comes stealing through the bush, till he observes me, and hastily departs.

After an hour of patient waiting the sound of clashing boughs is heard in the tree-tops. A few minutes later may be seen a big school of monkeys, led by a solemn-looking old pilot, who doubtless knows every palm that bears nuts within twenty miles around. They are now coming to inspect my cage, and see what new thing this is, set up in monkeydom.

As they come nearer, they become more cautious and tardy. They find a strong bough in the top of a big tree, and the grave old pilot perches himself far out on it, to peep at my cage. Just behind him sits the next in rank, resting his hands on the shoulders of the leader, while a dozen more are arranged in similar attitudes behind each other, along the limb. Each one pushes the one just in front of him, to make him move up a little closer, but no one of them, except the pilot, seems to want the front seat.

They look in silence, turning their little heads from side to side, as if to be certain it is not an illusion. They nudge one another again, and move up an inch or two closer, squinting their bright eyes, as if in doubt about the strange sight before them. They have made such calls before, but have not quite determined what kind of an animal this is in the cage. At each successive visit they come a little nearer, until now they are not a hundred feet away. Now they take alarm at something, and hurry away in another direction.

Next comes an armadillo, prowling about for insects among the leaves. He catches a glimpse of the cage, he stands motionless for a moment, to see what it is, and then, like a flash, he is gone.

During this time birds of divers kinds have been flying in all directions. Some of them perch on the limbs near by, some pick the nuts of the palm-tree, while others scream and screech, like so many tin-whistles, or brass horns. Many of them are parrots. Some have brilliant and beautiful plumage.

It is now ten o'clock. Not a breath of air stirs a leaf of the whole forest. The heat is sweltering and oppressive. The voices of the birds grow less and less frequent. Even the insects do not appear to be so busy as they were in the earlier hours of the day. Moses has abandoned his rambles in the bush, and sits on a fallen tree, with his arms folded, as if he had finished work for the day.

Along towards this hour everything in the forest appears to become quiet and inactive, and continues so until about two o'clock in the afternoon. I was impressed on more than one occasion with this universal rest during the hottest part of the day, and the same thing seems to prevail among the aquatic animals.

I now prepare my repast for midday, by opening a can of meat or fish, and warming it in a tin plate on the little stove. I have no vegetables or dessert, but with a few crackers broken up, and stirred into the grease, and plenty of water to drink with it, I find it an ample meal. When it is finished, Moses coils up in his little hammock, swung by my side, and takes his siesta. The boy, when there, stretches out on the floor, and does likewise.

During the hours from ten till two, few things are astir, though I have seen some interesting sights during that time.

It must not be supposed that the change is sudden at these periods, for such is not the case. It is not a fixed time for everything to cease its activity. It is by slow degrees that one after another becomes quiescent, until life appears almost extinct for a time; but as the sun begins to descend the western sky, things begin to revive, and by three o'clock everything is again astir.

Now a lone gorilla comes stalking through the bush, looking for the red fruit of the *batuna* that grows at the root of the plant. He plucks a bud of some kind, tears it apart with his fingers, smells it, and throws it aside. Now he takes hold of a tall sapling, looks up at the shaking branches, and turns aside. He pauses and looks around as if suspicious of danger. He listens to see if anything is approaching, but being reassured he resumes his search for food. Now he gently parts the tangled vines that intercept his way, and creeps noiselessly through them. He hesitates, looks carefully around him, and then proceeds again. He is coming this way. I can see his black face as he turns his head from side to side, looking for food. What a brutal visage! It has a scowl upon it, as if he were at odds with all his race. He is now within a few yards of the cage, but is not aware of my presence. He plucks the tendril from a vine, smells it, and puts it in his mouth. He plucks another and another. I shall note that vine, and ascertain what it is. Now he is in a small open space, where the bush is cut away, so as to afford a better view. He seems to know that this is an unusual thing to find in the jungle, so he surveys it with caution. He comes nearer. Now he has detected me. He sits down upon the ground, and looks at me as if in utter surprise. A moment more he turns aside, looks back over his shoulders, but hurries away into the dense jungle.

It is now four o'clock, and I hear a wild pig rooting among the fallen leaves. I see a small rodent that looks like a diminutive hedgehog. He is gnawing the bark from a dead limb, possibly to capture some insect secreted under it; but as rodents usually live upon vegetable diet, he may have some other reason for this.

It is five o'clock, and the shadows are beginning to deepen in the forest. I see two little grey monkeys playing in the top of a very tall tree. The birds are tiresome and monotonous. Yonder is a small snake twined around the limb of a bushy tree. He is doubtless hunting for a nest of young birds. The low, muttering sound of distant thunder is heard, but little by little it grows louder. It is the familiar voice of the tornado. I must prepare for it.

The stove is now lighted, and a pie-pan of water set on it. In it is stirred an ounce of desiccated soup. It is heated to the boiling-point, and then set on the swinging table. Then a can of mutton is emptied into another pan of the same kind, and a few crackers broken and stirred in. The soup is eaten while the meat is being cooked. When it is ready, the flame of the stove is turned off, and the second course of dinner is served, consisting of canned mutton, crackers and water. The dishes, consisting usually of three tin pie-pans and a cup, are thrust out into the adjacent bush, for the ants and other insects to clean during the night.

In the meantime Moses has had his supper, and gone to his own little cage, to find shelter from the approaching storm. The curtains are hung up on the side of the cage, from which the tornado is coming. Now the leaves begin to rustle. It is the first cool breath of the day, but it is only the herald of the furious wind that is rapidly advancing. The tree-tops begin to sway. Now they are lashing each other as if in anger; the strong trees are bending from the wind; the lightning is so vivid that it is blinding; the thunder is terrific. One shaft after another, the burning bolts are hurled through the moaning forest. The roar of thunder is unceasing. I hear the dull thud of a falling tree, while the crackling boughs are falling all around me. The rain is pouring in torrents, and all nature is in a rage. Every bird and beast has sought a place of refuge from the warring elements. No sign of life is visible, no sound is audible, save the voice of the storm.

How unspeakably desolate the jungle is at such an hour, no fancy can depict. How utterly helpless a human being is against the wrath of nature, no one can realise, except to live through such an hour in such a place.

PREPARING FOR THE NIGHT

On one occasion five large trees were blown down, within a radius of two hundred yards of my cage, and scores of limbs were broken off by the wind, and scattered like straws. Some of them were six or eight inches in diameter, and ten or twelve feet long. One of them broke the corner off the bamboo roof over my cage. The limb was broken off a huge cotton-tree near by, and fell from a height of about sixty feet. It was carried by the wind some yards out of a vertical line as it fell, and just passed far enough to spare my cage. Had it struck the body of it, no doubt it would have been partly demolished, for the main body of the bough was about six inches in diameter and ten feet long. This particular tornado lasted for nearly three hours, and was the most violent of any I saw during the entire year.

Now the storm subsides, but the darkness is impenetrable. I have no light of any kind, for that would alarm the inhabitants of the jungle, and attract a vast army of insects from all quarters. Moses and the boy are fast asleep, while I sit and listen to the many strange and weird sounds heard in the jungle at night The bush crackles near by. It is a leopard creeping through it. He is coming this way. Slowly, cautiously he approaches. I cannot see him in the deep shadows of the foliage, but I can locate him by sound, and identify him by his peculiar tread. Perhaps he will attack the cage when he gets near enough. He is creeping up closer. He evidently smells his prey, and is bent on seizing it.

My rifle stands by my elbow. I silently raise it, and lay it across my lap. The brute is now crouching within a few yards of me, but I cannot see to shoot him. I hear him move again, as if adjusting himself to spring upon the cage. He cannot see it, but he has located me by scent. I hear a low rustling of the leaves as he wags his tail preparatory to a leap. If I could only touch a button and turn on a bright electric light over his head! He remains crouching near, while I sit with the muzzle of my rifle turned towards him, and my hand on the lock. It is a trying moment. If he should spring with such force as to break the frail network that is between us, there could be but one fate for me.

In the brief space of a few seconds a thousand things run through one's mind. Not prompted by fear, but by suspense. Is it best to fire into the black shadows, or to wait for his attack? What is his exact pose? What does he intend? How big is he? Can he see me? And a category of similar questions arise at this critical moment.

A clash of bushes, and he is gone. Not with the stealthy, cautious steps with which he advanced, but in hot haste. He has taken alarm, abandoned his purpose, and far away I can hear the dry twigs crashing as he hurries to some remote nook. He flees as if he thought he was being pursued. He is gone, and I feel a sense of relief.

It is ten o'clock, the low rumbling of distant thunder is all that remains of the tornado that swept over me a few hours ago. The stars are shining, but the foliage of the forest is so dense, that I can only see one here and there, peeping through the tangled boughs overhead. I hear some little waif among the dead leaves, but what it is, or what it wants, can only be surmised.

Another hour is passed, and I retire to my hammock. The sounds of nocturnal birds are fewer now. I hear a strange, tremulous sound up in the boughs of the bushes near the cage. It sounds like the leaves vibrating. It ceases, and begins again at intervals. I listen with attention, for it is very singular. It is a huge python in search of birds. He reaches his head and neck forward, grasps the bough of a slender bush, releases his coil from another, and by contraction draws his slimy body forward. The pliant boughs yield to his heavy weight. The abrasion causes it to tremble, and the leaves to quake.

I fall asleep and rest in comfort, while the dew that has fallen on the leaves gathers itself into huge drops, their weight bends the leaves, and they fall from their lofty perch, striking those far below with a sharp, popping sound. The hours fly by, but in the stillness of the early morning is heard a most unearthly scream. It is a king gorilla. He simply makes every leaf in the forest tremble with the sound of his piercing shrieks.

The dawn again awakes to life the teeming forest, and all its denizens again go forth to join the universal chase for food.

All of these incidents cited are true in every detail, but they did not occur every day, nor did all of them occur on the same day, as would be inferred from the manner in which they are related.

This gives a glimpse of my real daily life in the jungle, but the monotony was often relieved by going out for a day or two at a time, or hunting on the plains, a few miles away. My menu was occasionally varied by a chicken, piece of goat, fish or porcupine; but the general average of it was about as described.

CHAPTER IV
THE CHIMPANZEE

Next to man, the chimpanzee occupies the highest plane in the scale of nature. His mental and social traits, together with his physical type, assign him to this place.

In his distribution, he is confined to Equatorial Africa. His habitat, roughly outlined, is from the fourth parallel north of the equator to the fifth parallel south of it, along the west coast, and extends eastward about half-way across the continent. His range can be defined with more precision, but its exact limits are not quite certain. Its boundary on the north is defined by the Kameroon valley, slightly curving to the north, but its extent eastward is not well known. He does not appear to be found anywhere north of this river, and it is quite certain that the few specimens attributed to the north coast of the Gulf of Guinea do not belong to that territory. On the south, its boundary starts from the coast, at a point near the fifth parallel, curves northward, crossing the Congo near Stanley Pool, pursues a north-east course, to the centre of the Congo State, again curves southward, across the Upper Congo, towards the north end of Lake Tanganyika. Its limits appear to conform more to isothermal lines, than to the rigid lines of geometry.

Specimens are sometimes secured by collectors beyond the limits mentioned, but so far as I can ascertain they appear to have been captured within these limits. There are numerous centres of population. This ape is not strictly confined to any definite topography, but occupies the upland forests or the low basin lands.

In one section he is known to the natives by one name, and in another by quite a different one. The name *chimpanzee* is of native origin. In the Fiot tongue the name of the ape is *chimpan*, which is a slight corruption of the true name. It is properly a compound word, the first syllable is from the Fiot word *tyi*, which white men erroneously pronounce like "chee." It means "small," and is found in many of the native compounds. The latter syllable is from *mpâ*, a bushman, hence the word literally means, in the Fiot tongue, "a small bushman."

Among other tribes the common name of the ape is *ntyigo*. The two names appear to come from the same ultimate source. The latter is derived from the Mpongwe word *ntyia*, blood, hence breed, and the word *iga*, the forest, and literally means the "breed of the forest." The same idea is involved in the two names, and both convey the oblique idea that the animal is something more like man than other animals are.

There are two distinct types of this ape, and they are now regarded as two species. One of them is distributed throughout the entire habitat described, while the other is only known south of the equator, between the second and fifth parallels, and west of the Congo. Both kinds are found within these limits, but the variety which is confined to that region is called, by the tribes that know the ape, the *kulu-kamba*, in contradistinction to the other kind, known as *ntyigo*. This name is derived from *kulu*, the onomotope of the sound made by the animal and the native verb *kamba*, to speak, hence the name literally means the thing "that speaks kulu."

In certain points the common variety differs from the *kulu-kamba* in a degree that would indicate that they belong to distinct species, but the skulls and skeletons are so nearly the same, that no one can identify them with certainty. In life, however, it is not difficult to distinguish them.

The *ntyigo* has a longer face and more prominent nose than the *kulu*. His complexion is of all shades of brown, from a light tan to a dark, dingy mummy colour. He has a thin coat of short black hair, which is often described as brown, but that effect is due to the colour of his skin blending with that of his suit. In early life his hair is quite black, but in advanced age the ends are tipped with a dull white, giving him a dingy grey colour. The change is due to the same causes that produce grey hairs on the human body. But there is one point in which they differ. The entire hair of the human becomes white with age, while only the end of it does so in the chimpanzee. In the human, one hair becomes white, while another retains its natural colour, but in this ape all the hairs appear to undergo the same change.

In very aged specimens the outer part of the hair often assumes a dirty, brownish colour, which is due to the want of vascular action to supply the colour pigment, and the same effect is often seen in preserved specimens, for the same reason that the hair of an Egyptian mummy is brown, while in life it was doubtless a jet black. In this ape the hair is uniformly black, except the small tuft of white at the base of the spinal column and a few white hairs on the lower lip and chin. I have examined about sixty living specimens and I have never found any other colour among them only from the cause mentioned. The normal colour of both sexes is the same.

The *kulu*, as a rule, has but little hair on the top of its head, but that on the back of it and on the neck is much longer than elsewhere on the body, and longer on them than on other apes.

Much stress is laid by some writers on the bald head of one ape and the parted hair on that of another. These features cannot be relied upon as having any specific meaning, unless there are as many species as there are apes. Sometimes a specimen has no hair on the summit of its head, while another differs from it in this respect alone by having a suit of hair more or less dense, and yet in every other respect they are the same. Some of them have the hair growing almost down to the eyebrows, and each hair appears to diverge from a common centre like the radii of a sphere: another of the same species will have the hair parted in the middle as neatly as if it had been combed, while another may have it in wild disorder. The same thing is noticed in certain monkeys, and it is equally true of the human being. As a factor in classifying them it signifies nothing. It may be remarked that as a whole the *kulu* is inclined to have little hair upon the crown of the head.

Between the two species there is a close alliance, but the males differ more than the females. This is especially true in the structure of certain organs.

The face in youth is quite free from hairs, but in the adult state there is, in both sexes, a slight tendency to grow a light down over the cheeks.

The colour of the skin is not uniform in all parts of the body, especially on the face. Some specimens have patches of dark colour set in a lighter ground. Sometimes certain parts of the face will be dark, and other parts light. I have seen one specimen quite freckled.

It is said by some that the skin is light in colour when young, and becomes darker with age, but such is not the case. It is true that the skin darkens a few shades as the cuticle hardens, but there is no transition from one colour to another, and this slight change of shade is only on the exposed parts.

The *kulu* has a short, round face, very much like that of a human. In early life it is quite free from hairs, but, like the other, a slight down appears with age. He has a heavy suit of hair on the body. It is coarser than that of the *ntyigo*, longer, and inclined to wave, giving it a fluffy aspect. The colour is jet black, except a small tuft of white about the base of the spine.

The skin varies in colour less than in the *ntyigo*, and the darker shades seldom appear. The eyes are a shade darker, and in both species the parts of the eye which are white in man are brown in the chimpanzee, gradually shading off into a yellow near the base of the optic nerve. As a rule, the *kulu* has a clear, open visage, with a kindly expression. It is confiding and affectionate to a degree beyond any other animal. It is more intelligent than its *confrère*, and displays the faculty of reason almost like a human being.

One important point in which these apes differ is in the scope and quality of voice. The *kulu* makes a greater range of vocal sounds than the other. Some of them are soft and musical, while those uttered by the *ntyigo* are fewer in number and more harsh in quality. One of them resembles the bark of a dog, and another is a sharp screaming sound.

The *kulu* evinces a certain sense of gratitude, while the *ntyigo* appears to be almost devoid of this instinct. There are many traits in which they differ, but human beings, of the same family, also differ in these qualities.

The points in which they coincide are many, and after a brief review of them, we may consider the question of making two species of them, or assigning them to the same.

The skeletons, as we have noted, are the same in form, size and proportion. Their muscular, nervous, and veinous systems are the same, except a slight structural variation in the genital organs of the males, and the degree of mobility in certain facial muscles. The character of their food, and the mode of eating it, are the same in each. In captivity they appear to regard each other as one of their own kind, but whether they mate or not remains to be learned.

Such is the sum of the likenesses and differences between the two extreme types of this genus; but with so many points in common, and so few in which they differ, it is a matter of serious doubt whether they can be said to constitute two distinct species, or only two marked varieties of a common species. This doubt is further emphasised by the fact that all the way between these two extremes are many gradations of intermediate types, so that it is next to impossible to say where one ends and the other begins.

In view of all these facts, I believe them to be two well-defined varieties of the same species; they are the white man and the negro of a common stock. They are the patrician and plebeian of one race, or the nobility and yeomanry of one tribe. They are like different phases of the same moon. The *kulu-kamba* is simply a high order of chimpanzee.

IN THE JUNGLE

It is quite true that two varieties of one species usually have the same vocal characteristics, and this appears to be the strongest point in favour of assigning them to separate species, but it is not impossible that even this may be waived.

Leaving this question for others to decide, as they find the evidence to sustain them, we shall, for the present, regard them as one kind, and consider their physical, social and mental status.

Whether they be all of one species, or divided into many, the same habits, traits, and modes of life prevail throughout the entire group, so that one description will apply to all, so far as we have to deal with them in general. There are many incidents to be related elsewhere, which apply to individuals of the special kinds mentioned, but for the present the term chimpanzee is meant to include the whole group, except where it may be otherwise specified.

CHAPTER V
PHYSICAL, SOCIAL, AND MENTAL QUALITIES

Physically, the chimpanzee, as we have seen, closely resembles man, but there are certain points that have not been mentioned in which he differs from him, also from other apes. We may here take note of a few of those points.

The model and structure of the ear of this ape are somewhat the same as those of man, but the organ is larger in size, and thinner in proportion. It is very sensitive to sound, but dull to the touch, indicating that the surface is not well provided with nerves. He cannot move it as other animals move theirs by the use of the muscles at its base, but, like the human ear, it is quite fixed and helpless in this respect.

The hand of the chimpanzee is long and narrow. The finger bones are longer, in proportion to their size, than those of the human hand, and slightly more curved in the plane of the digits. One thing peculiar in the hand of the chimpanzee, is that the tendons inside of the hand, which are called the flexors, and designed to close the fingers, are shorter than the line of the bones, and on this account the fingers of the ape are always held in a curve, so that he cannot possibly straighten them into a line. This is probably due to the habit of climbing in which he indulges to a great extent; also to the practice of hanging by the hands. In making his way through the bush, he often swings from bough to bough by the arms alone, and sometimes suspends himself by one arm, while he uses the other to pluck and eat fruit. This characteristic is transmitted to the young, and is found in the first stages of infancy. The thumb is not truly opposable, but is inclined to close towards the palm of the hand. It is of little use to him. His nails are thick, dark in colour, and not so flat as those of man.

Instead of having the great toe in line with the others, it projects at an angle from the side of the foot, something after the manner of the human thumb. The foot itself is flexible, and has great prehensile power. In climbing, and in many other ways, it is used as a hand. The tendons in the sole of the foot are equal in length to the line of the bones, and the digits of the foot can be straightened, but both members are inclined to curve into an arch in the line of the first and second digits.

His habit of walking is peculiar. The greater part of the weight is borne upon the legs. The sole of the foot is placed almost flat on the ground, but the pressure is greatest along the outer edge of it, in the line of the last digit. This is easily noticed where he walks through plastic ground. In the act of walking he always uses the hands, but does not place the palm on the ground; he uses the backs of the fingers instead, sometimes only the first joints are placed on the ground, resting on the nails; at other times the first and second joints are used, while at others the backs of all the fingers from the knuckles to the nails serve as a base for the arm. The integument on these parts is not callous, like that of the palm; the colour pigment is distributed the same as on other exposed parts of the body, which shows that the weight of the body is not borne on the fore limbs, as it is in the case of a true quadruped, but indicates that the hand is only used to balance the body and shift the weight from foot to foot, while in the act of walking. The weight is not equally distributed between the hands and the feet.

His waddling gait is caused by his short legs, stooping habit and heavy body. All bipeds with stout bodies and short legs are predisposed to a waddling motion, which is due to the wide angle between the weight and the changing centre of gravity.

The chimpanzee is neither a true quadruped, nor a true biped, but combines the habits of both. It appears to be a transition state from the former to the latter, and a vestige of this habit is still to be found in man, whose arms alternate in motion with his legs in the act of walking, which suggests the idea that he may, at some time, have had a similar habit of locomotion. Such a fact does not show that he was ever an ape, but it does point to the belief that he has once occupied a like horizon in nature to that now occupied by the ape, and that having emerged from it, he still retains traces of the habit.

This peculiarity is still more easily observed in children than in adults. In early infancy all children are inclined to be bow-legged, and in their first efforts at walking, invariably press most of their weight on the outer edge of the foot, and curve the toes inward, as if to grasp the surface on which the foot is placed. The instinct to prehension cannot be mistaken; it differs in degree in different races, and is vastly more pronounced in negro than in white infants.

There is another peculiar feature in the walk of the chimpanzee. The motion of the arms and legs do not alternate with the same degree of regularity that they do in man or quadrupeds. This ape uses his arms more like crutches. They are moved forward, not quite, but almost at the same instant, and the motion of the legs is not at equal intervals. To be more explicit: the hands are placed almost opposite each other; the right foot is advanced about three times its length; the left foot placed about one length in front of it; the arms are again moved; the right foot again advanced about three lengths forward of the left; and the left again brought about one length in front of it. The same animal does not always use the same foot to make the long stride. It will be seen by this that each foot moves through the same space, and that in a line, the tracks of either foot are the same distance apart, but the distance from the track of the right foot to that of the left is about three times as great as the distance from the track of the left foot to that of the right; or the reverse may be the case. The distance from the track of either foot to the succeeding track of the other, is never the same between the right and left tracks, except where the animal is walking at great leisure.

There is, perhaps, no animal more awkward than the chimpanzee, when he attempts to run. He sometimes swings his body with such force between his arms as to lose his balance, and falls backward on the ground. I have often seen him do this, and when he would right himself again, would be half his length farther backward than forward of his starting-point.

The chimpanzee is doubtless a better climber than the gorilla. He finds much of his food in trees, but is not arboreal in habit in the proper sense of that term. To be arboreal, the animal must sleep in trees or on a perch, but the chimpanzee cannot do so. He sleeps the same as a human being does. He lies down on the back or side, and, as a rule, uses his arms for a pillow. I do not believe it possible for him to sleep on a perch. He may sometimes doze in that way, but the grasp of his foot is only brought into use when he is conscious of it. I have often known Moses to climb down from the trees and lie upon the ground to take a nap. I never even saw him so much as doze in any other position.

I may here call attention to one fact concerning the arboreal habit. There appears to be a rule to which this habit conforms. Among apes and monkeys the habit is in keeping with the size of the animal. The largest monkeys, as a rule, are only found among the lowest trees, and the smaller monkeys among the taller trees. It is a rare thing ever to see a large monkey in the top of a tall tree. He may venture there for food or to make his escape, but it is not his proper element. This same rule appears to hold good among the apes themselves. The gibbon has this habit in a more pronounced degree than any other true ape. The orang appears to be next; the chimpanzee then comes in for a third place, and the gorilla last. It must not be understood that all of these apes do not frequently climb, even to the tops of the highest trees; but that is not their normal mode of life any more than the top of a mast is the proper place on a ship for a sailor.

The chimpanzee is nomadic in habit, and, like the gorilla, seldom or never passes two nights in the same spot. As to his building huts or nests in trees or elsewhere, I am not prepared to believe that he ever does so. I hunted in vain, for months, and made diligent inquiry in several tribes, but failed to find a specimen of any kind of shelter built by an ape. I do not assert that it is absolutely untrue, but I have never been able to obtain any evidence, except the statement of the natives that it was true. On the contrary, certain facts point to the opposite belief. If the ape built him a permanent home the natives would soon discover it, and there would be no difficulty in having it pointed out. If he built a new one every night, however rude and primitive it might be there would be so many of them in the forest that there would be no difficulty in finding them. The nomadic habit plainly shows that he does not build the former kind, and the utter absence of them shows that he does not build the latter kind, and the whole story appears to be without foundation.

In addition to these facts, one thing to be noticed is that few or none of the mammals of the tropics ever build any kind of a home. Even the animals that have the habit of burrowing in other climates, do not appear to do so in the tropics. This is due, no doubt, to the warm climate, in which they are not in need of shelter. Of course birds, and other oviperous animals, build nests, as they do elsewhere.

The longevity of these apes is largely a matter of conjecture, but from a cursory study of their dentition and other factors of their development, it appears that the male reaches the adult stage at an age ranging from nine to eleven years, while the female matures at six or seven. These appear to be the periods at which they pass from the state of adolescence. Some of them live to be perhaps forty years of age, or upwards, but the average of life is doubtless not more than twenty-two or twenty-three years. The average of life is more uniform with them than with man. These figures are not mere guesswork, but are deduced from reliable data.

The period of gestation in both these apes is a matter that cannot be stated with certainty. Some of the natives say that it is nine months, while others believe that it is seven months or less, and there are some facts to support both of these claims, but nothing quite conclusive. The sum of the evidence that I could find rather pointed to a term of three months or thereabouts as the true period. During the months of February and March the male gorillas are vociferous in their screaming, the young adults separate from the families, and some other things indicate that this is the season of pairing and breeding. Such may not be the case, but the inference is well-founded. It is quite certain that the season of bearing the young is from the beginning of May to the end of June. It is about this time that the dry season begins and continues for four months. It would appear that nature has selected this period of the year because it is more favourable for rearing the young. During this season food is more abundant and can be secured with less effort. The lowlands are drier, and this enables the mother to retire to the dense jungle with her young, where she is less exposed to danger than she would be in the more open forest.

It is not certain whether the periods are the same with both apes or not, and native reports differ on this point, but it is probable that they are the same.

From a social point of view, the chimpanzee appears to be of a little higher caste than other animals. In his marital ideas he is polygamous, but is, in a certain degree, loyal to his family. The paternal instinct is a trifle more refined in him than in most other animals. He seems to appreciate the relationship of parent and child more, and retain it longer than others do. Most male animals discard their young, and become estranged to them at a very early age; but the chimpanzee keeps his children with him until they are old enough to go away and rear a family of their own.

The family of the chimpanzee frequently consists of three or four wives and ten or twelve children, with one adult male; but there are cases known in which two or three elderly males have been seen in the same family, but they appear to have their own wives and children. In such an event, however, there seems to be one who is supreme. This fact suggests the idea that among them a form of patriarchal government prevails. The wives and children do not appear to question the authority of the patriarch, or to rebel against it. The male parent often plays with his children, and appears to be fond of them.

A STROLL IN THE JUNGLE

There is one universal error that I desire here to correct. It is the common idea that animals are so strongly possessed of the parental instinct that they nobly sacrifice their own lives in defence of their young. I do not wish to dispel any belief that tends to dignify or ennoble animals, for I am their special friend and champion; but truth demands that we qualify this statement. It is quite true that many have lost their lives in such acts of defence, but it was not a voluntary sacrifice. It was not alone in the defence of their young, but in many cases it was in self-defence. In others, it was from a lack of judgment. These apes have often been frightened away from their young, and the latter captured while the parents were fleeing from the scene. This may have been the result of sagacity rather than of depravity, but the parental instinct in both sexes, in many instances, has failed to restrain them from flight. If it be a foe that appears to come within the measure of their own power, they will certainly defend their young, and this sometimes results in the loss of their own lives; but if it be one of such formidable aspect as to appear quite invincible, the parents leave the young to their fate. This is true of many other animals, including man.

I have no desire to detract from the heroic quality of this instinct, or to dim the glory it sheds upon noble deeds ascribed to it; but the fact that a parent incurs the risk of its own life in the defence of its young, is not a true test of its strength or quality. It is only in the few isolated cases of a voluntary sacrifice of the parent, foreknowing the result, that it can be said the act was due to the instinct. In most cases it is under the belief in its ability to rescue the one in danger, but the parent is not wholly aware of its own danger.

I doubt if any animal except man ever deliberately offered its own life as a ransom for that of another, and such instances in human history are so rare as to immortalise the actor.

To whatever extent the instinct may be found, it is much stronger in the female than in the male, and it appears to be stronger in domestic animals than in wild ones. To what extent this is due to their contact with man, it is difficult to say. The germ may be inherent, but it certainly yields to culture.

The fact of the ape deserting its offspring under certain conditions, may be taken as an evidence of its superior intelligence and its appreciation of life and danger, rather than a low, brutish impulse. It is the exercise of superior judgment that causes man to act with more prudence than other animals. It does not detract from his nobleness.

Within the family circle of the chimpanzee the father is supreme; but he does not degrade his royalty by being a tyrant. Each member of the family seems to have certain rights that are not impugned by others. For example, possession is the right of ownership. When one ape procures a certain article of food, the others do not try to dispossess it. It is from this source, doubtless, that man inherits the idea of private ownership. It is the same principle amplified by which nations hold the right of territory, but nations often violate this right, and so do chimpanzees when not held in check by something more potent than a sense of justice. With all due respect, I do not think the ape abuses the right by urging his claim beyond his real needs, while nations sometimes do.

When a member of a family of apes is ill, the others are quite conscious of it, and evince a certain amount of solicitude. Their conduct indicates that they have, in a small degree, the passion of sympathy, but the emotion is feeble and wavering. So far as I know, they do not essay any treatment, except to soothe and comfort the sufferer. They surely have some definite idea of what death is, and I have reason to believe that they have a name for it. They do not readily abandon their sick, but when one of them is unable to travel with the band, the others rove about for some days, within call of it, but do not minister to its wants.

It is said, if one of them is wounded, the others will rescue it if possible, and convey it to a place of safety; but I cannot vouch for this, as such an incident has never come within my own experience.

One of the most remarkable of all the social habits of the chimpanzee, is the *kanjo*, as it is called in the native tongue. The word does not mean "dance" in the sense of saltatory gyrations, but implies more the idea of "carnival." It is believed that more than one family takes part in these festivities.

Here and there in the jungle is found a small spot of sonorous earth. It is irregular in shape, but is about two feet across. The surface is of clay, and is artificial. It is superimposed upon a kind of peat bed, which, being very porous, acts as a resonance cavity, and intensifies the sound. This constitutes a kind of drum. It yields rather a dead sound, but of considerable volume.

This queer drum is made by chimpanzees, who secure the clay along the bank of some stream in the vicinity. They carry it by hand, and deposit it while in a plastic state, spread it over the place selected, and let it dry. I have, in my possession, a part of one that I brought home with me from the Nkami forest. It shows the finger-prints of the apes, which were impressed in it while the mud was yet soft.

After the drum is quite dry, the chimpanzees assemble by night in great numbers, and the carnival begins. One or two will beat violently on this dry clay, while others jump up and down in a wild and grotesque manner. Some of them utter long, rolling sounds, as if trying to sing. When one tires of beating the drum, another relieves him, and the festivities continue in this fashion for hours.

I know of nothing like this in the social economy of any other animal, but what it signifies, or what its origin was, is quite beyond my knowledge. It appears probable that they do not indulge in this *kanjo* in all parts of their domain, nor do they occur at regular intervals.

The chimpanzee is averse to solitude. He is fond of the society of man, and is easily domesticated. If allowed to go at liberty, he is well-disposed, and is strongly attached to man, but if confined, he becomes vicious and ill-tempered. All animals, including man, have the same tendency.

Mentally the chimpanzee occupies a high plane within his own sphere of life, but within those limits the faculties of the mind are not called into frequent exercise, and therefore they are not so active as they are in man.

It is difficult to compare the mental status of the ape to that of man, because there is no common basis upon which the two rest. Their modes of life are so unlike, as to afford no common unit of measure. Their faculties are developed along different lines. The two have but few problems in common to solve. While the scope of the human mind is vastly wider than that of the ape, it does not follow that it can act with more precision in all things. There are, perhaps, instances in which the mind of the ape excels that of man, by reason of its adaptation to certain conditions. It is not a safe and infallible guide to measure all things by the standard of man's opinion of himself. It is quite true that, by such a unit of measure, the comparison is much in favour of the man, but the conclusion is neither just nor adequate.

It is a problem of great interest, however, to compare them in this manner, and the result would indicate that a fair specimen of the ape is in about the same mental horizon as a child of one year old. But if the operation were reversed, and man were placed under the natural conditions of the ape, the comparison would be much less in his favour. There is no common mental unit between them.

The chimpanzee exercises the faculty of reason with a fair degree of precision, on problems that concern his own comfort or safety. He is quick to interpret motives, to discern intents, and is a rare judge of character. He is inquisitive, but not so imitative as monkeys are. He is more observant of the relations of cause and effect, and in his actions he is controlled by more definite motives. He is docile, and quickly learns anything that lies within the range of his own mental plane.

The opinion has long prevailed that these apes subsist upon a vegetable diet, but such is not in anywise the case. In this respect their habits are the same as those of man, except that the latter has learned to cook his food, while the former eats his raw.

Their natural tastes are much diversified, and they are not all equally fond of the same articles of food. Most of them are partial to the wild mango, which grows in abundance in certain localities in the forest, and is often available when other kinds of food are scarce. It thus becomes, as it were, a staple article of food. There are many kinds of nuts to be found in their domain, but the oil palm nut appears to be a favourite. They also eat the kola nut, when it is to be had. Several kinds of small fruits and berries also form a part of their diet. They eat the stalks of some plants, the tender buds of others, and the tendrils of certain vines, the names of which I do not know.

Most of the fruits and plants that are relished by them are either acidulous or bitter in taste, and they are not especially fond of sweet fruits, if they can get those having the flavours mentioned. They eat bananas, pine-apples, and other sweet fruits, but not from choice. Most of them appear to prefer a lime to an orange, a plantain to a banana, or a kola nut to a sweet mango, but in captivity they acquire a taste for sweet foods of all kinds.

In addition to these articles they devour birds, lizards, and small rodents. They rob the birds of their eggs and their young. They make havoc on many kinds of large insects. Those that I have owned were fond of cooked meats and salt fish, either raw or cooked.

CHAPTER VI
THE SPEECH OF CHIMPANZEES

The speech of chimpanzees is limited to a few sounds, and these are confined chiefly to their natural wants. The entire vocabulary of their language embraces perhaps not more than twenty words, and many of them are vague or ambiguous, but they express the concept of the ape with as much precision as it is defined to his mind, and quite distinctly enough for his purpose.

In my researches I have learned about ten words of his speech, so that I can understand them, and make myself understood by them. Most of these sounds are within the compass of the human voice, in tone, pitch, and modulation; but two of them are much greater in volume than it is possible for the human lungs to reach, and one of them rises to a pitch more than an octave higher than any human voice. These two sounds are audible at a great distance, but they do not fall within the true limits of speech.

THE EDGE OF THE JUNGLE

The vocal organs of this ape resemble those of man as closely as any other character has been shown to resemble. They differ slightly in one detail that is worthy of notice. Just above the opening called the glottis, which is between the vocal cords, are two small sacs or ventricles. These, in the ape, are larger and more flexible than in man. In the act of speaking they are inflated by the air passing out of the lungs through the long tube called the larynx. The function of these organs is to control and modify the sound by increasing or decreasing the pressure of the air that is jetted through this tube. They serve, at the same time, as a reservoir and a gauge.

In the louder sounds produced by the chimpanzee these ventricles distend until the membrane of which they are composed is held at a high tension. This greatly intensifies the voice, and increases its volume. It is partly due to these little sacs that the ape is able to make such a loud and piercing scream. But the pitch and volume of his voice cannot be due to this cause alone, for the gorilla, in which these ventricles are much smaller, can make a vastly louder sound, unless we are mistaken about the one ascribed to him.

Although the sounds made by the chimpanzee can be imitated by the human voice, they cannot be expressed or represented by any system of phonetic symbols in use among men. All alphabets have been deduced from pictographs, and the symbol that represents any given sound has no reference to the organs that produced it. The few rigid lines that have survived to form the alphabets are conventional, and within themselves meaningless, but they have been so long used to represent these sounds of speech that it would be difficult to supplant them with others, even if such were desired.

As no literal formula can be made to represent the phonetic elements of the speech of chimpanzees, I have taken a new step in the art of writing by framing a system of my own, which is rational in plan and simple in device.

The organs of speech always act in harmony, and a certain movement of the lips is always attended by a certain movement of the internal organs of speech. This is true of the ape as well as of man, and in order to utter the same sounds each would employ the same organs, and use them in the same way.

By this means, deaf mutes are able to distinguish the sounds of speech and reproduce them, although they do not hear them. By close study and long practice they learn to distinguish the most delicate shades of sound.

In this plain fact lies the clue to the method I have used. It is, as yet, only in the infant state, but it is possible to be made, with a very few symbols, to represent the whole range of vocal sounds made by man or other animals.

The chief symbols I employ are the parentheses used in common print. The two curved lines placed with the convex sides opposite, thus, (), represent the open glottis, in which position the voice will utter the deep sound of "O." The glottis about half closed utters the sound of "U," as in the German, and to represent this sound a period is inserted between the two curved lines, thus, (.). When the aperture is contracted still more it produces the sound of "A" broad, and to represent this a colon is placed between the lines, thus, (:). When the aperture is restricted to a still smaller compass the sound of "U" short is uttered, and to represent this an apostrophe is placed between the lines, thus, ('). When the vocal cords are brought to a greater tension, and the aperture is almost closed, it utters the short sound of "E." To represent this sound a hyphen is inserted between the lines, thus, (-). These are the main vowel sounds of all animals, although in man they are sometimes modified, and to them is added the sound of "E" long, while in the ape the long sounds of "O" and "E" are rarely, if ever, heard.

From this vowel basis all other sounds may be deduced, and by the use of diacritics to indicate the movement of the organs of speech, the consonant elements may be easily expressed.

A single parenthesis, with the concave side to the left, will represent the initial sound of "W," which seldom, but sometimes, occurs in the sounds of animals. When used, it is placed on the left side of the leading symbol, thus,)(), and this symbol, as it stands, should be pronounced nearly like "U-O," but with the first letter suppressed, and almost inaudible. Turning the concave side to the right, and placing it on the right side of the symbol, it represents the vanishing sound of "W," thus, ()(. This symbol reads "O-U," with the "O" long, and the "W" depressed into the short sound of "U." The apostrophe placed before or after the symbol will represent "F" or "V." The grave accent, thus, (`), represents the breathing sound of "H," whether placed before or after the symbol, and the acute accent, thus, ('), will represent the aspirate sound of that letter in the same way.

When the symbol is written with a numeral exponent, it indicates the degree of loudness. If there is no figure, the sound is such as would be made by the human voice in ordinary speech. The letter "X" will indicate a repetition of the sound, and the numeral placed after it will show the number of times repeated, instead of the degree of loudness. For example, we will write the sound (.), which is equivalent to long "U," made in a normal tone, the same symbol written thus (.)2 indicates the sound, made with greater energy, and about twice as loud. To write it thus, (.)X2, indicates that the sound was repeated, and so on.

One peculiar sound made by these animals, which is described in connection with the gorilla, appears to be the result of inhalation, but I know of no other animal that makes a sound in this manner.

As an example of the use of this method, we will write the French word "feu," which Moses mastered, thus, '(), which is equivalent to "vû" with the

"U" sounded short, the other word "wie," in German, thus,)('), which is pronounced almost like "wû," giving "u" the short sound again.

I shall not lead the reader through the long and painful task by giving the entire system as far as I have gone, but what has been given will convey an idea of a system, by means of which it will be possible to represent the sounds of all animals, so that the student of phonetics will recognise at once the character of the sound, even if he cannot reproduce it by natural means.

It would be tedious and of no avail to the casual reader to reduce to writing here the sounds made by the chimpanzee; but it may be of interest to mention and describe the character and use of some of them.

Perhaps the most frequent sound made by all animals, appears to be that referring to food, and therefore it may claim the first place in our attention. This word in the language of the chimpanzee begins with the short sound of the vowel "u" which blends into a strong breathing sound of "h," the lips are compressed at the sides, and the aperture of the mouth is nearly round. It is not difficult to imitate, and the ape readily understands it even when poorly made.

Another sound of frequent use among them is that used for calling. The vowel element is nearly the same, though slightly sharpened, and merges into a distinct vanishing "w." The food sound is often repeated two or three times in succession, but the call is rarely ever repeated, except at long intervals.

One sound is particularly soft and musical, the vowel element is that of long "u" as in the German. This blends into a "w," followed by the slightest suggestion of the short sound of "a." It appears to express affection or love. This sound is also the first of the series of sounds attributed to the gorilla.

The most complex sound made by them is the one elsewhere described as meaning "good." They often use it in a sense very much the same as mankind uses the word "thanks," but it is not probable that they use it as a polite term, yet the same idea is present.

One of the words of warning or alarm contains a vowel element closely resembling the short sound of "e." It terminates with the breathing sound of "h." It is used to announce the approach of anything that he is familiar with, and not afraid of. If the sound is intended to warn against the approach of an enemy, or something strange, the same vowel element is used, but terminates with the aspirate sound of "h" pronounced with energy and distinctness. The two words are the same in vowel quality, but they differ in the time required to utter them, and the final breathing and aspirate effects. There is also a difference in the manner of the speaker in the act of delivering the word, which plainly indicates that he knows the use and value of the sounds. At the approach of danger the latter is often given almost in a whisper, and at long intervals apart, but increases in loudness as the danger approaches; the other is usually spoken distinctly and repeated frequently. It is worthy of note that the native tribes often use the same word in the same manner and for the same purpose.

There are other sounds which are easily identified but difficult to describe, such as that used to signify "cold" or "discomfort"; another for "drink"; another referring to "illness," and still another which I have good reason to believe means "dead" or "death." There are perhaps a dozen more that I can distinguish, but have not yet been able to determine their meaning. I have an opinion as to some of them which I have not yet verified.

The chimpanzee makes use of a few signs which seem to be fixed factors of expression. He makes a negative sign by moving the head from side to side, but the gesture is not frequent or pronounced. Another negative sign, which is more common, is a motion of the hand from the body towards the person or thing addressed. This sign is sometimes made with great emphasis, and there can be no question as to what it means. The manner of making the sign is not uniform. Sometimes it is done by an urgent motion of the hand. Bringing it from his opposite side, with the back forward, it is waved towards any one approaching, if the ape object to the approach. The same sign is often made as a refusal of anything offered him. Another way of making this sign is with the arm extended forward, the hand hanging down, and the back towards the person approaching or the thing refused. In addition to these negative signs there is one which may be regarded as affirmative. It is made simply by extending one arm towards the person or thing desired. It sometimes serves the purpose of beckoning; but in this act there is no motion of the hand. These signs are similar in character to those used by men, and appear to be innate.

It must not be inferred from this small list of words and signs that there is nothing left to learn. So far we have only taken the first step as it were in the study of the speech of apes. As we grow more familiar with their sounds, it becomes less difficult to understand them. I have not been disappointed in what I hoped to learn from these animals. The total number of words in the speech of all simians that I have learned up to this time is about one hundred. I have given no attention of late to the small monkeys, but I shall resume the task at some future day, as it forms a part of the work I have assumed, but all of that is described in a work already published.

In conclusion, I will say that the sounds uttered by these apes have all the characteristics of true speech. The speaker is conscious of the meaning of the sound used, and uses it with the definite purpose of conveying an idea to the one addressed; the sound is always addressed to some definite one, and the speaker usually looks at the one addressed; he regulates the pitch and volume of the voice to suit the condition under which it is used; he knows the value of sound as a medium of thought. These and many other facts show that they are truly speech.

If these apes were placed under domestication, and kept there as long as the dog has been, he would be as far superior to the dog in sagacity as he is by nature above the wild progenitors of the canine race.

CHAPTER VII
THE CAPTURE AND CHARACTER OF MOSES

During my sojourn in the forest, I had a fine, young chimpanzee, which was of ordinary intelligence, and of more than ordinary interest, because of his history.

I gave him the name Moses, not in derision of the historic Israelite of that name, but because of the circumstances of his capture and life.

He was found all alone in a wild papyrus swamp of the Ogowe River. No one knew who his parents were, or how he ever came to be left in that dismal place. The low bush in which he was crouched when discovered was surrounded by water, and the poor little waif was cut off from the adjacent dry land.

As the native who captured him approached, the timid little ape tried to climb up among the vines above him, and escape, but the agile hunter seized him before he could do so. At first the chimpanzee screamed, and struggled to get away, because he had perhaps never before seen a man, but when he found that he was not going to be hurt, he put his frail arms around his captor, and clung to him as a friend. Indeed, he seemed glad to be rescued from such a dreary place, even by such a strange creature as a man.

For a moment the man feared that the cries of his young prisoner might call its mother to the rescue, and possibly a band of others; but if she heard them, she did not respond, so he tied the baby captive with a thong of bark, put him into his canoe, and brought him away to the village, where he supplied him with food, and made him quite cosy. The next day he was sold to a trader. About this time I passed up the river on my way to the jungle in search of the gorilla and other apes. Stopping at the station of the trader, I bought him, and took him along with me. We soon became the best of friends and constant companions.

It was supposed that the mother chimpanzee left her babe in the tree while she went off in search of food, and wandered so far away that she lost her bearings and could not again find him. He appeared to have been for a long time without food, and may have been crouching there in the forks of that tree for a day or two; but such was only inferred from his hunger, as there was no way to determine how long he had remained, or even how he got there.

I designed to bring Moses up in the way that good chimpanzees ought to be brought up, so I began to teach him good manners in the hope that some day he would be a shining light to his race, and aid me in my work among them. To that end I took great care of him, and devoted much time to the study of his natural manners, and to improving them as much as his nature would allow.

I built him a neat little house within a few feet of my cage. It was enclosed with a thin cloth, and had a curtain hung at the door, to keep out mosquitoes and other insects. It was supplied with plenty of soft, clean leaves, and some canvas bed-clothing. It was covered over with a bamboo roof, and suspended a few feet from the ground, so as to keep out the ants.

Moses soon learned to adjust the curtain, and go to bed without my aid. He would lie in bed in the morning until he heard me or the boy stirring about the cage, when he would poke his little black head out, and begin to jabber for his breakfast. Then he would climb out, and come to the cage to see what was going on.

He was not confined at all, but quite at liberty to go about in the forest, climb the trees and bushes, and have a good time of it. He was jealous of the boy, and the boy was jealous of him, especially when it came to a question of eating. Neither of them seemed to want the other to eat anything that they mutually liked, and I had to act as umpire in many of their disputes on that grave subject, which seemed to be the central thought of both of them.

I frequently allowed Moses to dine with me, and I never knew him to refuse, or to be late in coming on such occasions, but his table etiquette was not of the best order. I gave him a tin plate and a wooden spoon, but he did not like to use the latter, and seemed to think that it was pure affectation for any one to eat with such an awkward thing. He always held it in one hand, while he ate with the other, or drank his soup out of the plate.

It was such a task to get washing done in that part of the world, that I resorted to all means of economy in that matter, and for a tablecloth I used a leaf of newspaper, when I had it. To tear that paper afforded Moses an amount of pleasure that nothing else would, and in this act his conduct was more like that of a naughty child than in anything he did.

When he would first take his place at the table, he behaved in a nice and becoming manner; but having eaten till he was quite satisfied, he usually became rude and saucy. He would slily put his foot up over the edge of the table, and catch hold of the corner of the paper, meanwhile watching me closely, to see if I was going to scold him. If I remained quiet he would tear it just a little and wait to see the result. If no notice was taken of that, he would tear it a little more, but keep watching my face to see when I observed it. If I raised my finger to him, he quickly let go, drew his foot down, and began to eat. If nothing more was done to stop him, the instant my finger and eyes were dropped, that dexterous foot was back on the table and the mischief resumed with more audacity than before.

When he carried his fun too far, I made him get down from the table and sit on the floor. This humiliation he did not like at best, but when the boy would grin at him for it, he would resent it with as much temper as if he had been poked with a stick. He certainly was sensitive on this point, and evinced an undoubted dislike to being laughed at.

Another habit that Moses had was putting his fingers in the dish to help himself. He had to be watched all the time to prevent this, and seemed unable to grasp any reason why he should not be allowed to do so. He always appeared to think my spoon, knife and fork were better than his own spoon. On one occasion he persisted in begging for my fork until I gave it to him. He dipped it into his soup, held it up, and looked at it as if disappointed. He again stuck it into his soup, and then examined it, as if to see how I lifted my food with it. He did not seem to notice that I used it in lifting meat instead of soup. After repeating this three or four times, he licked the fork, smelt it, and then deliberately threw it on the floor, as if to say, "That's a failure." He leaned over and drank his soup from the plate.

The only thing that he cared much to play with was a tin can that I kept some nails in. For this he had a kind of mania, and never tired of trying to remove the lid. When given the hammer and a nail, he knew what they were for, and would set to work to drive the nail into the floor of the cage or the table; but he hurt his fingers a few times, and after that he stood the nail on its flat head, removed his fingers and struck it with the hammer, but, of course, never succeeded in driving it into anything.

A bunch of sugar-cane was kept for Moses to eat when he wanted it, and to aid him in tearing the hard shell away from it, I kept a club to bruise it. Sometimes he would go and select a stalk of the cane, carry it to the block, take the club in both hands, and try to mash the cane himself; but as the jar of the stroke often hurt his hands, he learned to avoid this, by letting go as the club descended. He never succeeded in crushing the cane, but would continue his efforts until some one came to his aid. At other times he would drag a stalk of the cane to the cage, poke it through the wires, then bring the club, and poke it through, to get me to mash it for him.

From time to time I received newspapers sent me from home. Moses could not understand what induced me to sit holding that thing before me, but he wished to try it, and see. He would take a leaf of it, and hold it up before him with both hands, just as he saw me do; but instead of looking at the paper, he kept his eyes, most of the time, on me. When I would turn mine over, he did the same thing, but half the time had it upside down. He did not appear to care for the pictures, or notice them, except a few times he tried to pick them off the paper; and one large cut of a dog's head, when held at a short distance from him, he appeared to regard with a little interest, as if he recognised it as that of an animal of some kind, but I cannot say just what his ideas concerning it really were.

Chimpanzees are not usually so playful or funny as monkeys, but they have a certain degree of mirth in their nature, and at times display a marked sense of humour.

One thing that Moses liked was to play peek-a-boo with me or the boy. He did not try to conceal his body from view, but would hide his eyes, and then peep. A favourite time for this was in the early part of the afternoon. He would often go and put his head behind a large tin box in the cage, while his whole body was visible. In this attitude he would utter a series of peculiar sounds, then draw his head out, and look at me, to see if I was watching him. If not, he would repeat the act a few times, and then hunt something else to amuse himself with. But if he could gain attention, the romp began, and he found great pleasure in this simple pastime. He would roll over, kick up his heels, and grin, with evident delight.

I spent much time in entertaining him in this way, and felt amply repaid for it in the gratification it afforded him. I could not resist his overtures to play, as he was my companion and my friend, and, living in that solitary gloom, it was a mutual pleasure.

Another occasion on which he used to peep at me was when he lay down to take his midday nap. For this I had made him a little hammock, which was suspended by wires, so that it could be removed when not in use. I always hung this by my side in the cage, so I could swing him to sleep like a child. He liked this, and I liked to indulge him. When he was laid in it, he was usually covered up with a small piece of canvas, and in spreading it over him, I frequently laid the edge of it over his eyes, but he seemed to suspect me of having some motive in doing so. Often he would reach his fingers up, catch the edge of the cloth, and gently draw it down, so he could see what I was doing. If he saw that he was detected, he would quickly release it, and cuddle down, as if it had been done by accident; but the little rogue knew, just as well as I did, what it meant to peep.

I also made him another hammock, and hung it out a few yards from the cage, so he could get into it without bothering me; but he never cared for it, until I brought a young gorilla to live with us in our jungle home, and as Moses never used it, I assigned it to the new member of the household. Whenever the gorilla got into it there was a small row about it. Moses would never allow him to occupy it in peace. He seemed to know that it was his own by right, and the gorilla was regarded as an intruder. He would push and shove the gorilla, grunt and whine and quarrel, until he got him out of it; but after doing so he would leave it, and climb up into a bush, or go away to hunt something to eat. He only wanted to dispossess the intruder, for whom he nursed an inordinate jealousy. He never went near the gorilla's little house, which was on the opposite side of the cage from his own; even after the gorilla died, he kept aloof from it.

As a rule, I always took Moses with me in my rambles into the forest, and I found him to be quite useful in one way. His eyes were like the lens of a camera—nothing escaped them; and when he discovered anything in the jungle, he always made it known by a peculiar sound. He could not point it out with his finger, but by watching his eyes the object could often be located.

Frequently during these tours the ape rode on my shoulders, and at other times the boy carried him, but occasionally he was put down on the ground to walk. If we travelled at a very slow pace, and allowed him to stroll along at leisure, he was content to do so, but if hurried beyond a certain gait he always made a display of his temper. He would turn on the boy and attack him, if possible; but if the boy escaped, the angry little ape would throw himself down on the ground, scream, kick, and beat the earth with his own head and hands in the most violent and persistent manner. He sometimes did the same way when not allowed to have what he wanted. His conduct was exactly like that of a spoiled, ugly child.

He had a certain amount of ingenuity, and often evinced a degree of reason which was rather unexpected. It was not a rare thing for him to solve some problem that involved a study of cause and effect, but always in a limited degree. I would not be understood to mean that he could work out any abstract problem, such as belongs to the realm of mathematics, but simple, concrete problems, where the object was present.

On one occasion, while walking through the forest we came to a small stream of water. The boy and myself stepped across it, leaving Moses to get over it without help. He disliked getting his feet wet, and paused to be lifted across. We walked a few steps away, and waited. He looked up and down the branch to see if there was any way to avoid it. He walked back and forth a few yards, but found no way to cross it. He sat down on the bank, and declined to wade it. After a few moments he waddled along the bank, about ten or twelve feet, to a clump of tall slender bushes growing by the edge of the stream. Here he halted, whined, and looked up into them thoughtfully. At length he began to climb one of them that leaned over the water. As he climbed up, the stalk bent with his weight, and in an instant he was swung safely across the little brook. He let go the plant, and came hobbling along to me with a look of triumph on his face that plainly indicated that he was fully conscious of having performed a very clever feat.

One dark, rainy night I felt something pulling at my blanket and mosquito bar. I could not for a moment imagine what it was, but knew that it was something on the outside of my cage. I lay for a few seconds, and felt another strong pull at them. In an instant some cold, damp, rough thing touched my face, and I found it was his hand poked through the meshes and groping about for something. I spoke to him, and he replied with a series of plaintive sounds which assured me that something must be wrong.

I arose, and lighted a candle. His little brown face was pressed up against the wires, and wore a sad, weary look. He could not tell me in words what troubled him, but every sign, look, and gesture bespoke trouble. Taking the candle in one hand, and my revolver in the other, I stepped out of the cage and went to his domicile, where I discovered that a colony of ants had invaded his quarters.

These ants are a great pest when they attack anything, and when they make a raid on a house the only thing to be done is to leave it until they have devoured everything about it that they can eat. When they leave a house there is not a roach, rat, bug, or insect left in it.

As the house of Moses was so small, it was not difficult to dispossess them by saturating it with kerosene, which was quickly done, and the little occupant allowed to return and go to bed. He watched the procedure with evident interest, and seemed perfectly aware that I could rid him of his savage assailants. In a wild state he would doubtless have abandoned his claim, and fled to some other place without an attempt to drive them away, but in this instance he had acquired the idea of the rights of possession.

Moses was especially fond of corned beef and sardines, and would recognise a can of either as far away as he could see it. He also knew the instrument used in opening them, but he did not appear to appreciate the fact that when the contents had once been taken out it was useless to open the can again, so he often brought the empty cans that had been thrown into the bush, would get the can-opener down, and want me to use it for him. I never saw him try to open it himself, except with his fingers. Sometimes, when about to prepare my own meals, I would open the case in which I kept stored a supply of canned meats, and allow Moses to select one for the purpose. He never failed to pull out one of the cans of beef, bearing the red and blue label. If I put it back he would select the same kind, and could not be deceived in his choice. It was not accidental, because he would hunt for one until he found it.

I don't know what he thought when it was not served for dinner, as I often exchanged it for another kind without consulting him.

I kept my supply of water in a large jug, which was placed in the shade of the bushes near the cage. I also kept a small pan for Moses to drink out of. He would sometimes ask for water, by using his own word for it. He would place his pan by the side of the jug and repeat the sound a few times.

If he was not attended to he proceeded to help himself. He could take the cork out of the jug quite as well as I could. He would then put his eye to the mouth of it, and look down into the vessel to see if there was any water. Of course the shadow of his head would darken the interior of the jug so that he could not see anything. Then removing his eye from the mouth of it, he would poke his hand in it, but I reproved him for this until I broke him of the habit. After a careful examination of the jug he would try to pour the water out. He knew how it ought to be done, but was not able to handle the vessel himself. He always placed the pan on the lower side of the jug; then leaned the jug towards it and let go. He would rarely ever get the water into the pan, but always turned the jug with the neck down grade. As a hydraulic engineer he was not a great success, but he certainly knew the first principles of the science.

I tried to teach Moses to be cleanly, but it was a hard task. He would listen to my precepts as if they had made a deep impression, but he would not wash his hands of his own accord. He would permit me or the boy to wash them, but when it came to taking a bath, or even wetting his face, he was a rank heretic on the subject, and no amount of logic would convince him that he needed it. When he was given a bath, he would scream and fight during the whole process; and when it was finished he would climb up on the roof of the cage and spread himself out in the sun. This was the only occasion on which I ever knew him to get up on the roof. I don't know why he disliked it so much. He did not mind getting wet in the rain, but rather seemed to like that.

He had a great dislike for ants and certain large bugs. Whenever one came near him he would talk like a magpie, and brush at it with his hands until he got rid of it. He always used a certain sound for this kind of annoyance; it differed slightly from those I have described as warning.

Moses tried to be honest, but he was affected with a species of kleptomania, and could not resist the temptation to purloin anything that came in his way. The small stove upon which I prepared my food was placed on a shelf in one corner of the cage, about half-way between the floor and the top. Whenever anything was set on the stove to cook, he had to be watched to keep him from climbing up the side of the cage, reaching his arm through the meshes and stealing it. He was sometimes very persevering in this matter. One day I set a tin can of water on the stove to heat in order to make some coffee; he silently climbed up, reached his hand through, stuck it in the can, and began to search for anything it might contain. I threw out the water, refilled the can, and drove him away. In a few minutes he returned and repeated the act. I had a piece of canvas hung up on the outside of the cage to keep him away. The can of water was placed on the stove for the third time, but within a minute he found his way by climbing up under the curtain between it and the cage. I determined to teach him a lesson. He was allowed to explore the can, but finding nothing he withdrew his hand, and sat there clinging to the side of the cage. Again he tried, but found nothing. The water was getting warmer, but was still not hot. At length, for the third or fourth time he stuck his hand into it up to the wrist. By this time the water was so hot that it scalded his hand. It was not severe enough to do him any harm, but quite enough so for a good lesson. He jerked his hand out with such violence that he threw the cup over, and spilt the water all over that side of the cage. From that time to the end of his life he always refused anything that had steam or smoke about it. If anything having steam or smoke was offered him at the table, he would climb down at once and retire from the scene. Poor little Moses! I knew beforehand what would happen, and I did not wish to see him hurt, but nothing else would serve to impress him with the danger and keep him out of mischief.

Anything that he saw me eat he never failed to beg. No matter what he had himself, he wanted to try everything else that he saw me eat. One thing in which these apes appear to be wiser than man is, that when they eat or drink enough to satisfy their wants they quit, while men sometimes do not. They never drink water or anything else during their meal, but, having finished it, as a rule they always want something to drink. The native custom is the same. I have never known the native African to use any kind of diet drink, but always when he has finished eating takes a draught of water.

Moses knew the use of nearly all the tools that I carried with me in the jungle. He could not use them for the purpose they were intended, and I do not know to what extent he appreciated their use, but he knew quite well the manner of using them. I have mentioned the incident of his using the hammer and nails, but he also knew the way to use the saw; however, he always applied the back of it, because the teeth were too rough, but he gave it the motion. When allowed to have it, he would put the back of it across a stick and saw with the energy of a man on a big salary. When given a file, he would file everything that came in his way; and if he had applied himself in learning to talk human speech as closely and with as much zeal as he tried to use my pliers, he would have succeeded in a very short time.

Whether these creatures are actuated by reason or by instinct in such acts as I have mentioned, the cavillist may settle for himself; but it accomplishes the purpose of the actor in a logical and practical manner, and they are perfectly conscious that it does.

CHAPTER VIII
THE LIFE AND DEATH OF MOSES

I know of nothing in the way of affection and loyalty among animals that can exceed that of my devoted Moses. Not only was he tame and tractable, but he never tired of caressing me, and being caressed by me. For hours together he would cling to my neck, play with my ears, lips and nose, bite my cheek, and hug me like a last hope. He was never willing for me to put him down from my lap, never willing for me to leave my cage without him, never willing for me to caress anything else but himself, and never willing for me to discontinue that. He would cry and fret for me whenever we were separated, and I must confess that my absence from him during a journey of three weeks, hastened his sad and untimely death.

From the second day after we became associated, he appeared to regard me as the one in authority. He would not resent anything I did to him. I could take his food out of his hands, which he would permit no one else to do. He would follow me, and cry after me like a child; and as time went by his attachment grew stronger and stronger. He gave every evidence of pleasure at my attentions, and evinced a certain degree of appreciation and gratitude in return. He would divide any morsel of food with me, which is, perhaps, the highest test of the affection of any animal. I cannot say that such an act was genuine benevolence, or an earnest of affection in a true sense of the term, but nothing except deep affection or abject fear impels such actions, and certainly fear was not his motive.

There were others whom he liked and made himself familiar with; there were some he feared and others he hated; but his manner towards me was that of deep affection. It was not alone in return for the food he received, because my boy gave him food more frequently than I did, and many others from time to time fed him. His attachment was like an infatuation that had no apparent motive, was unselfish and supreme.

The chief purpose of my living among the animals being to study the sounds they uttered, I gave strict attention to those made by Moses. For a time it was difficult to detect more than two or three distinct sounds, but as I grew more and more familiar with them I could detect a variety of them, and by constantly watching his actions and associating them with his sounds I learned to interpret certain ones to mean certain things.

In the course of my sojourn with him I learned a certain sound that he always uttered when he saw anything that he was familiar with, such as a man or a dog, but he could not tell me which of the two it was. If he saw anything strange to him he could tell me, but not so that I knew whether it was a snake or a leopard or a monkey, yet I knew that it was something of that kind. I learned a certain word for food, hunger, eating, &c., but he could not go into any details about it, except that a certain sound indicated good or satisfaction, and another meant the opposite.

Among the sounds that I learned was one that is used by a chimpanzee in calling another to come to it. Some of the natives assured me that the mothers always used it in calling their young to them. When Moses wandered away from the cage into the jungle, he would sometimes call me with this sound. I cannot express it in letters of the alphabet, nor describe it so as to give a very clear idea of its character. It was a single sound or word of one syllable, and easily imitated by the human voice. At any time that I wanted Moses to come to me I used this word, and the fact that he always obeyed it by coming confirmed my opinion as to its meaning. I do not think when he addressed it to me that he expected me to come to him, but he perhaps wanted to locate me in order to be guided back to the cage by the sound. As he grew more familiar with the surrounding forest he used it less frequently, but he always employed it in calling me or the boy. When he was called by it he answered with the same sound; but one fact that we noticed was that if he could see the one who called he never made any reply by sound. He would obey it, but not answer it; he probably thought if he could see the one who called that he could be seen by him, and it was therefore useless to reply.

The speech of these animals is very limited, but it is sufficient for their purpose. It is none the less real because of its being restricted, but it is more difficult for man to learn, because his modes of thought are so much more ample and distinct. Yet when one is reduced to the necessity of making his wants known in a strange tongue, he can express many things in a very few words. I have once been thrown among a tribe of whose language I knew less than fifty words, but with little difficulty I succeeded in conversing with them on two or three topics. Much depends upon necessity, and more upon practice. In talking to Moses I mostly used his own language, and was surprised at times to see how readily we understood each other. I could repeat about all the sounds he made except one or two, but I was not able in the time we were together to interpret all of them. These sounds were more than a mere series of grunts or whines, and he never confused them in their meaning. When any one of them was properly delivered to him, he clearly understood and acted upon it.

It was never any part of my purpose to teach a monkey to talk, but after I became familiar with the qualities and range of the voice of Moses, I determined to see if he might not be taught to speak a few simple words of human speech. To effect this in the easiest way and shortest time, I carefully observed the movements of his lips and vocal organs in order to select such words for him to try as were best adapted to his ability.

I selected the word *mamma*, which may almost be considered a universal word of human speech; the French word *feu*, fire; the German word *wie*, howl, and the native Nkami word *nkgwe*, mother. Every day I took him on my lap and tried to induce him to say one or more of these words. For a long time he made no effort to learn them, but after some weeks of persistent labour and a bribe of corned beef, he began to see dimly what I wanted him to do. The native word quoted is very similar to one of the sounds of his own speech, which means "good" or "satisfaction." The vowel element differs in them, and he was not able in the time he was under tuition to change them, but he distinguished them from other words.

In his attempt to say *mamma* he only worked his lips without making any sound, although he really tried to do so, and I believe that in the course of time he would have succeeded. He observed the movement of my lips, and tried to imitate them, but seemed to think that the lips alone produced the sound.

With *feu* he succeeded fairly well, except that the consonant element as he uttered it resembled "v" more than "f," so that the sound was more like *vu* making the u short as in "nut." It was quite as perfect as most people of other tongues ever learn to speak the same word in French, and if it had been uttered in a sentence, any one knowing that language would recognise it as meaning fire.

In his efforts to pronounce *wie* he always gave the vowel element like German "u" with the *umlaut*, but the "w" element was more like the English than the German sound of that letter.

Taking into consideration the fact that he was only a little more than a year old, and was in training less than three months, his progress was all that could have been desired, and vastly more than had been hoped for. Had he lived until this time, it is my belief that he would have mastered these and other words of human speech to the satisfaction of the most exacting linguist. If he had only learned one word in a whole lifetime, he would have shown at least that the race is capable of being improved and elevated in some degree.

Another experiment that I tried with him was one that I had used before in testing the ability of a monkey to distinguish forms. I cut a round hole in one end of a board and a square hole in the other, and made a block to fit into each one of them. The blocks were then given to him to see if he could fit them into the proper holes. After being shown a few times how to do this, he fitted them in without difficulty; but when he was not rewarded for the task by receiving a morsel of corned beef or a sardine, he did not care to work for the fun alone.

In colours he had but little choice, unless it was something to eat, but he could distinguish them with ease if the shades were pronounced.

I had no means of testing his taste for music or sense of musical sounds.

I must here take occasion to mention one incident in the life of Moses that never perhaps occurred before in the life of any other chimpanzee, and while it may not be of scientific value, it is at least amusing.

While living in the jungle, I received a letter enclosing a contract to be signed by myself and a witness. Having no means of finding a witness to sign the paper, I called Moses from the bushes, placed him at the table, gave him a pen and had him sign the document as witness. He did not write his name himself, as he had not yet mastered the art of writing, but he made his cross mark between the names, as many a good man had done before him. I wrote in the blank the name,

His
"MOSES X NTYIGO"
mark;

the cross mark omitted, and had him with his own hand make the cross as it is legally done by all people who cannot write. With this signature the contract was returned in good faith to stand the test of the law courts of civilisation, and thus for the first time in the history of the race a chimpanzee signed his name.

When I prepared to start on a journey across the Esyira country it was not practicable for me to take Moses along, so I arranged to leave him in charge of a missionary. Shortly after my departure the man was taken with fever, and the chimpanzee was left to the care of a native boy belonging to the mission. The little prisoner was kept confined by a small rope attached to his cage in order to keep him out of mischief. It was during the dry season, when the dews are heavy and the nights chilly, as the winds at that season are fresh and frequent.

Within a week after leaving him he contracted a severe cold, which soon developed into acute pulmonary troubles of a complex type, and he began to decline. After an absence of three weeks and three days, I returned to find him in a condition beyond the reach of treatment. He was emaciated to a living skeleton: his eyes were sunken deep into their orbits, and his steps were feeble and tottering; his voice was hoarse and piping; his appetite was gone, and he was utterly indifferent to anything around him.

During my journey I had secured a companion for him, and when I disembarked from the canoe, I hastened to him with this new addition to our little family. I had not been told that he was ill, and was not prepared to see him looking so ghastly.

When he discovered me approaching, he rose up and began to call me as he had been wont to do before I left him, but his weak voice was like a death-knell to my ears. My heart sunk within me as I saw him trying to reach out his long, bony arms to welcome my return. Poor, faithful Moses! I could not repress the tears of pity and regret at this sudden change, for to me it was the work of a moment. I had last seen him in the vigour of a strong and robust youth, but now I beheld him in the decrepitude of a feeble senility. What a transformation!

I diagnosed his case as well as I was able and began to treat him, but it was evident that he was too far gone to expect him to recover. My conscience smote me for having left him, yet I felt that I had not done wrong. It was not neglect or cruelty for me to leave him while I went in pursuit of the chief object of my search, and I had no cause to reproach myself for having done so. But emotions that are stirred by such incidents are not to be controlled by reason or hushed by argument, and the pain that it caused me was more than I can tell.

If I had done wrong, the only restitution possible for me to make was to nurse him patiently and tenderly to the end, or till health and strength should return. This was conscientiously done, and I have the comfort of knowing that the last sad days of his life were soothed by every care that kindness could suggest. Hour after hour during that time he lay silent and content upon my lap. That appeared to be a panacea to all his pains. He would roll his dark brown eyes up and look into my face, as if to be assured that I had been restored to him. With his long fingers he stroked my face as if to say that he was again happy. He took the medicines I gave him as if he knew their purpose and effect.

His suffering was not intense, but he bore it like a philosopher. He seemed to have some vague idea of his own condition, but I do not know that he foresaw the result. He lingered on from day to day for a whole week, slowly sinking and growing feebler, but his love for me was manifest to the last, and I dare confess that I returned it with all my heart.

Is it wrong that I should requite such devotion and fidelity with reciprocal emotion? No. I should not deserve the love of any creature if I were indifferent to the love of Moses. That affectionate little creature had lived with me in the dismal shadows of that primeval forest for so many long days and dreary nights; had romped and played with me when far away from the pleasures of home, and had been a constant friend alike through sunshine and storm. To say that I did not love him would be to confess myself an ingrate unworthy of my race.

The last spark of life passed away in the night. It was not attended by acute pain or struggling, but, falling into a deep and quiet sleep, he woke no more.

Moses will live in history. He deserves to do so, because he was the first of his race that ever spoke a word of human speech; because he was the first that ever conversed in his own language with a human being; and because he was the first that ever signed his name to any document; and Fame will not deny him a niche in her temple among the heroes who have led the races of the world.

CHAPTER IX
AARON

Having arranged my affairs in Fernan Vaz so as to make a journey across the great forest that lies to the south of the Nkami country and separates it from that of the Esyira tribe, I set out by canoe to a point on the Rembo about three days from the place where I had so long lived in my cage. At a village called Tyimba I disembarked, and after a journey of five days and a delay of three more days caused by an attack of fever, I arrived at a trading station near the head of a small river called Ndogo. It empties into the sea at Sette Kama, about four degrees south of the equator. The trading post is about a hundred miles inland, at a native village called Ntyi-ne-nye-ni, which, strange to say, means in the native tongue, "Some other place."

TRADING STATION IN THE INTERIOR

About the time I reached here, two Esyira hunters came from a distant village, and brought with them a smart young chimpanzee of the kind known in that country as the *kulu-kamba*. He was quite the finest specimen of his race that I have ever seen. His frank, open countenance, big brown eyes and shapely physique, free from mark or blemish of any kind, would attract the notice of any one who was not absolutely stupid.

It is not derogatory to the memory of Moses that I should say this, nor does it lessen my affection for him. Our passions are not moved by visible forces nor measured by fixed units: they disdain all laws of logic, and spurn the narrow bounds of reason; they obey no code of ethics that can be defined, and conform to no theory of action.

As soon as I saw this little ape I expressed a desire to own him, so the trader in charge bought him and presented him to me. As it was intended that he should be the friend and ally of Moses, although not his brother, we conferred upon him the name of Aaron, as the two names are so intimately associated in history that the mention of one always suggests the other.

Aaron was captured in the Esyira jungle by these same hunters, about one day's journey from the place where I secured him; and in this event began a series of sad scenes in the brief but varied life of this little hero that seldom come within the experience of any creature.

At the time of his capture his mother was killed in the act of defending him from the cruel hunters, and when she fell to the earth, mortally wounded, this brave little fellow stood by her trembling body, defending it against her slayers, until he was overcome by superior force, seized by his captors, bound with strips of bark, and carried away into captivity.

No human can refrain from admiring his conduct in this act, whether it was prompted by the instinct of self-preservation or by a sentiment of loyalty to his mother, for he was exercising that prime law of nature which actuates all creatures to defend themselves against attack, and his wild, young heart throbbed with like sensations to those of a human under a like ordeal.

I do not wish to appear sentimental by offering a rebuke to those who indulge in the sport of hunting, but much cruelty could be obviated without losing any of the pleasure of the hunt, and I have always made it a rule to spare the mother with her young. Whether animals feel the same degree of mental and physical pain as man or not, they do, in these tragic moments, evince a certain amount of concern for one another, which imparts a tinge of sympathy that must appeal to any one who is not devoid of every sense of mercy.

It is true that it is often difficult, and sometimes impossible, to secure the young by other means; but the manner of getting them often mars the pleasure of having them, and while Aaron was, to me, a charming pet and a valuable subject for study, I confess the story of his capture always touched me in a tender spot.

I may here mention that the few chimpanzees that reach the civilised parts of the world are but a small percentage of the great number that are captured. Some die on their way to the coast, others die after reaching it, and scores of them die on board the ships to which they are consigned for various ports of Europe and other countries. It is not often from neglect or cruelty, but usually from a change of food, climate, or condition, yet the creature suffers just the same whether the cause is from design or accident. One fruitful source of death among them is pulmonary trouble of various types.

One look at the portrait of Aaron will impress any one with the high mental qualities of this little captive, but to see and study him in life would convince a heretic of his superior character. In every look and gesture there was a touch of the human that no one could fail to observe. The range of facial expression surpassed that of any other animal I have ever studied. In repose, his quaint face wore a look of wisdom becoming to a sage; while in play it was crowned with a grin of genuine mirth. The deep, searching look he gave to a stranger was a study for the psychologist, while the serious, earnest look of inquiry when he was perplexed would amuse a stoic. All these changing moods were depicted in his mobile face, with such intensity as to leave no room to doubt the activity of certain faculties of the mind in a degree far beyond that of animals in general; and his conduct, in many instances, showed the exercise of mental powers of a higher order than that limited agency known as instinct.

In addition to these facts, his voice was of better quality and more flexible than that of any other specimen I have ever known. It was clear and smooth in uttering sounds of any pitch within its scope, while the voices of most of them are inclined to be harsh or husky, especially in sounds of high pitch.

Before leaving the village where I secured him, I made a kind of sling for him to be carried in. It consisted of a short canvas sack with two holes cut in the bottom for his legs to pass through. To the top of this was attached a broad band of the same cloth by which to hang it over the head of the carrier boy to whom the little prisoner was consigned. This afforded the ape a comfortable seat, and at the same time reduced the labour of carrying him. It left his arms and legs free, so he could change his position and rest, while it also allowed the boy the use of his own hands in passing any difficult place in the jungle along the way.

PLAIN AND EDGE OF THE FOREST

From there to the Rembo was a journey of five days on foot. Along the way were a few straggling villages, but most of the route lay through a wild and desolate forest, traversed by low broad marshes, through which wind shallow sloughs of filthy greenish water, seeking its way among bending roots and fallen leaves. From the foul bosom of these marshes rise the effluvia of decaying plants, breeding pestilence and death. Here and there across the dreary tracts is found the trail of elephants, where the great beasts have broken their tortuous way through the dense barriers of bush and vine. These trails serve as roads for the native traveller, and afford the only way of crossing these otherwise trackless jungles.

The only means of passing these dismal swamps is to wade through the thin slimy mud, often more than knee-deep, and sometimes extending many hundred feet in width, intercepted at almost every step by the tangled roots of mangrove-trees under foot, or clusters of vines hanging from the boughs overhead.

Such was the route we came, but Aaron did not realise how severe the task of his carrier was in trudging his way through such places, and the little rogue often added to the labour by seizing hold of limbs or vines that hung within his reach in passing, and thus retarded the progress of the boy, who strongly protested against the ape amusing himself in this manner. The latter seemed to know of no reason why he should not do so, and the former did not deign to give one, and so the quarrel went on until we reached the river, but by that time each of them had imbibed a hatred for the other that nothing in the future ever allayed. Neither of them ever forgot it while they were associated, and both of them evinced their aversion on all occasions. The boy gave vent to his dislike by making ugly faces at the ape, which the latter resented by screaming and trying to bite him. Aaron refused to eat any food given him by the boy, and the boy would not give him a morsel except when required to do so. At times the feud became ridiculous, and it only ended in their final separation. The last time I ever saw the boy I asked him if he wanted to go with me to my country to take care of Aaron, but he shook his head, and said, "He's a bad man."

This was the only person for whom I ever knew Aaron to conceive a deep and bitter dislike, but the boy he hated with his whole heart.

On my return to Fernan Vaz, where I had left Moses, I found him in a feeble state of health as related elsewhere. When Aaron was set down before him, he merely gave the little stranger a casual glance, but held out his long lean arms for me to take him in mine. His wish was gratified, and I indulged him in a long stroll. When we returned I set him down by the side of his new friend, who evinced every sign of pleasure and interest. He was like a small boy when there is a new baby in the house. He cuddled up close to Moses and made many overtures to become friends, but while the latter did not repel them he treated them with indifference. Aaron tried in many ways to attract his attention, or to elicit some sign of approval, but it was in vain.

No doubt the manners of Moses were due to his health, and Aaron seemed to realise it. He sat for a long time, holding a banana in his hand, and looking with evident concern into the face of his little sick cousin. At length he lifted the fruit to the lips of the invalid and uttered a low sound, but the kindness was not accepted. The act was purely one of his own volition, in which he was not prompted by any suggestion from others, and every look and motion indicated a desire to relieve or comfort his friend. His manner was gentle and humane, and his face was an image of pity.

Failing to get any sign of attention from Moses, he moved up closer to his side and put his arms around him in the same manner that he is seen in the picture with Elisheba.

During the days that followed, he sat hour after hour in this same attitude, and refused to allow any one except myself to touch his patient; but on my approach he always resigned him to me, while he watched with interest to see what I did for him.

Among other things, I gave him a tabloid of quinine and iron twice a day. These were dissolved in a little water and given to him in a small tin cup which was kept for the purpose. When not in use, it was hung upon a tall post. Aaron soon learned to know the use of it, and whenever I would go to Moses, he would climb up the post and bring me the cup to administer the medicine.

It is not to be inferred that he knew anything about the nature or effect of the medicine, but he knew the use, and the only use, to which that cup was put.

During the act of administering the medicine, Aaron displayed a marked interest in the matter, and seemed to realise that it was intended for the good of the patient. He would sit close up to one side of the sick one and watch every movement of his face, as if to see what effect was being produced, while the changing expressions of his own visage plainly showed that he was not passive to the actions of the patient.

While I was present with the sick one, Aaron appeared to feel a certain sense of relief from the care of him, and frequently went climbing about as if to rest and recreate himself by a change of routine. While I would take Moses for a walk, or sit with him on my lap, his little nurse was perfectly content; but the instant they were left alone, Aaron would again fold him in his arms as if he felt it a duty to do so.

It was only natural that Moses, in such a state of health, should be cross and peevish at times, as people in a like condition are; but during the time I never once saw Aaron resent anything he did, or display the least ill-temper towards him, but, on the contrary, his conduct was so patient and forbearing that it was hard to forego the belief that it was prompted by the same motives of kindness and sympathy that move the human heart to deeds of tenderness and mercy.

At night, when they were put to rest, they lay cuddled up in each other's arms, and in the morning they were always found in the same close embrace; but on the morning Moses died, the conduct of Aaron was unlike anything I had observed before. When I approached their snug little house and drew aside the curtain, I found him sitting in one corner of the cage. His face wore a look of concern as if he was aware that something awful had occurred. When I opened the door, he neither moved nor uttered any sound. I do not know whether or not they have any name for death, but they surely know what it is.

Moses was dead. His cold body lay in its usual place, but was entirely covered over with the piece of canvas kept in the cage for bed-clothing. I do not know whether Aaron had covered him up or not, but he seemed to realise the situation. I took him by the hand and lifted him out of the cage, but he was reluctant. I had the body removed and placed on a bench about thirty feet away, in order to dissect and prepare the skin and skeleton to preserve them. When I proceeded to do this, I had Aaron confined to the cage, lest he should annoy and hinder me at the work; but he cried and fretted until he was released.

It is not meant that he wept or shed tears over the loss of his companion, for the lachrymal glands and ducts are not developed in these apes; but they manifest concern and regret which are motives of the passion of sorrow, but being left alone was the cause of this.

When released, he came and took his seat near the dead body, where he sat the whole day long and watched the operation.

After this he was never quiet for a moment if he could see or hear me, until I secured another of his kind for a companion; then his interest in me abated in a measure, but his affection for me remained intact.

His conduct towards Moses always impressed me with the belief that he appreciated the fact that he was in distress or pain, and while he may not have foreseen the result, he certainly knew what death was when he saw it. Whether it is instinct or reason that causes man to shrink from death, the same influence works to the same end in the ape; and the demeanour of this same ape towards his later companion, Elisheba, only confirmed the opinion.

CHAPTER X
AARON AND ELISHEBA

Four days after the death of Moses I secured a passage on a trading-boat that came into the lake. It was a small affair, intended for towing canoes, and not in any way prepared to carry passengers or cargo; but I found room in one of the canoes to set the cage I had provided for Aaron, stowed the rest of my effects wherever space permitted, and embarked for the coast.

Our progress was slow and the journey tedious, as the only passage out of the lake at that season was through a long, narrow, winding creek, beset by sand-bars, rocks, logs, and snags, and in some places overhung by low, bending trees. But the wild, weird scenery was grand and beautiful. Long lines of bamboo, broken here and there by groups of pendanus or stately palms; islands of lilies and long sweeps of papyrus, spreading away from the banks on either side; the gorgeous foliage of aquatic plants drooping along the margin like a massive fringe, and relieved by clumps of tall, waving grass, formed a perfect Eden for the birds and monkeys that dwell among those scenes of an eternal summer.

After a delay of eight days at Cape Lopez, we secured passage on a small French gunboat, called the *Komo*, by which we came to Gaboon, where I found another *kulu-kamba* in the hands of a generous friend, Mr. Adolph Strohm, who presented her to me; and I gave her to Aaron as a wife, and called her Elisheba, after the name of the wife of the great high-priest.

Elisheba was captured on the head-waters of the Mguni river, in about the same latitude that Aaron was found in, but more than a hundred miles to the east of that point and a few minutes north of it. I did not learn the history of her capture.

It would be difficult to find any two human beings more unlike in taste and temperament than these two apes were. Aaron was one of the most amiable of creatures; he was affectionate and faithful to those who treated him kindly; he was merry and playful by nature, and often evinced a marked sense of humour; he was fond of human society, and strongly averse to solitude or confinement.

Elisheba was a perfect shrew, and often reminded me of certain women that I have seen who had soured on the world. She was treacherous, ungrateful, and cruel in every thought and act; she was utterly devoid of affection; she was selfish, sullen, and morose at all times; she was often vicious and always obstinate; she was indifferent to caresses, and quite as well content when alone as in the best of company.

A NATIVE CANOE

It is true that she was in poor health, and had been badly treated before she fell into my hands, but she was by nature endowed with a bad temper and depraved instincts.

It is not at all rare to see a vast difference of manners, intelligence, and temperament among specimens that belong to one species. In these respects they vary as much in proportion to their mental scope as human beings do; but I have never seen, in any two apes of the same species, the two extremes so widely removed from one another.

While waiting at Gaboon for a steamer I had my own cage erected for them to live in, as it was large and gave them ample room for play and exercise. In one corner of it was suspended a small, cosy house for them to sleep in. It was furnished with a good supply of clean straw and some pieces of canvas for bed-clothes. In the centre of the cage was a swing, or trapeze, for them to use at their pleasure.

Aaron found this a means of amusement, and often indulged in a series of gymnastics that would evoke the envy of the king of athletic sports. Elisheba had no taste for such pastime, but her depravity could never resist the impulse to interrupt him in his jolly exercise. She would climb up and contend for possession of the swing until she would drive him away, when she would perch herself on it and sit there for a time in stolid content, but would neither swing nor play.

Frequently, when Aaron would lie down quietly on the straw during the day, she would go into the snug little house and raise a row with him by pulling the straw from under him, handful at a time, and throwing it out of the box till there was not one left in it.

No matter what kind or quantity of food was given them, she always wanted the piece he had, and would fuss with him to get it; but when she got it, she would sit holding it in her hand without eating it, for there were some things that he liked which she would not eat at all.

When we went out for a walk, no matter which way we started she always contended to go some other way; and if I yielded, she would again change her mind, and start off in some other direction. If forced to submit, she would scream and struggle as if for life.

I cannot forego the belief that these freaks were due to a base and perverse nature, and I could find no higher motive in her stubborn conduct.

Aaron was very fond of her, and rarely ever opposed her inflexible will. He clung to her, and let her lead the way. I have often felt vexed at him because he complied so readily with her wishes.

The only case in which he took sides against her was in her conduct towards me.

When I first secured her she had the temper of a demon, and with the smallest pretext she would assault me and try to bite me or tear my clothes. In these attacks Aaron was always with me, and the loyal little champion would fly at her in the greatest fury. He would strike her over the head and back with his hands, bite her, and flog her till she desisted. If she returned the blow he would grasp her hand and bite it, or strike her in the face. He would continue to fight till she submitted, when he would celebrate his victory by jumping up and down in a most grotesque fashion, stamping his feet, slapping his hands on the ground, and grinning like a mask. He seemed as conscious of what he had done and as proud of it as any human could have been; but no matter what she did to others, he was always on her side of the question. If any one else annoyed her, he would always resent it with violence.

About the premises there were natives all the time passing to and fro, and these two little captives were objects of special interest to them. They would stand by the cage hour after hour and watch them. The ruling impulse of nearly every native appears to be cruelty, and they cannot resist the temptation to tease and torture anything that is not able to retaliate. They were so persistent in poking my chimpanzees with sticks, that I had to keep a boy on watch all the time to prevent it; but the boy could not be trusted, so I had to watch him.

In the rear of the room that I occupied was a window through which I watched the boy and the natives both from time to time, and when anything went wrong I would call out from there to the boy. Aaron soon observed this, and found that he could get my attention himself by calling out when any one annoyed him, and he also knew that the boy was put there as a protector. Whenever any of the natives came about the cage he would call for me in his peculiar manner, which I well understood and promptly responded to. The boy also knew what it meant, and would rush to the rescue. If I were away from the house and the boy was aware of the fact, he was apt to be tardy in coming to the relief of the ape, and sometimes did not come at all, in which event the two would crawl into their house and pull down the curtain so that they could not be seen. Here they would remain until the natives would leave or some one came to their aid. Neither of them ever resented anything the natives did to them unless they could see me about, but whenever I came in sight they would make battle with their tormentors, and if liberated from the big cage, they would chase the last one of them out of the yard.

Aaron knew perfectly well that they were not allowed to molest him or his companion, and when he knew that he had my support he was ready to carry on the war to a finish. But it was really funny to see how meek and patient he was when left alone to defend himself against the natives with a stick, and then to note the change in him when he knew that he was backed up by a friend upon whom he could rely.

Mr. Strohm, the trader with whom I found hospitality at this place, kept a cow in the lot where the cage was. She was a small black animal, and the first that Aaron had ever seen. He never ceased to contemplate her with wonder and with fear. If she came near the cage when no one was about he hurried into his box, and from there peeped out in silence until she went away. The cow was equally amazed at the cage and its strange occupants, though less afraid, and frequently came near to inspect them. She would stand a few yards away with her head lifted high, her eyes arched and ears thrown forward, waiting for them to come out of that mysterious box; but they would not venture out of their asylum while she remained, until tired of waiting she would switch her tail, shake her head, and turn away.

When taken out of the cage, Aaron had special delight in driving the cow away, and if she was around he would grasp me by the hand and start towards her. He would stamp the ground with his foot, strike with all force with his long arm, slap the ground with his hand, and scream at her at the top of his voice. If she moved away, he would let go my hand and rush towards her as though he intended to tear her up; but if the cow turned suddenly towards him, the little fraud would run to me, grasp my leg, and scream with fright.

The cow was afraid of a man, and as long as she was followed by one she would continue to go; but when she would discover the ape to be alone in the pursuit, she would turn and look as if trying to determine what manner of thing it was. Elisheba never seemed to take any special notice of the cow except when she approached too near the cage, and then it was due to the conduct of Aaron that she made any fuss about it.

On board the steamer that we sailed in for home, there was a young elephant that was sent by a trader for sale. He was kept in a strong stall, built on deck for his quarters. There were wide cracks between the boards, and the elephant had the habit of reaching his trunk through them in search of anything he might find. With his long, flexible proboscis extended from the side of the stall, he would twist and coil it in all manner of writhing forms. This was the crowning terror of the lives of those two apes: it was the bogie-man of their existence, and nothing could induce either of them to go near it. If they saw me go about it, they would scream and yell until I came away. If Aaron could get hold of me without getting too near it, he would cling to me until he would almost tear my clothes to keep me away from it. It was the one thing that Elisheba was afraid of, and the only one against which she ever gave me warning.

They did not manifest the same concern for others, but sat watching them without offering any protest. Even the stowaway who fed them and attended to their cage was permitted to approach it, but their solicitude for me was remarked by every man on board.

I was never able to tell what their opinion was of the thing. They were much less afraid of the elephant when they could see all of him, than they were of the trunk when they saw that alone. They may have thought the latter to be a big snake, but such is only conjecture.

At the beginning of the voyage I took six panels of my own cage and made a small cage for them. I taught them to drink water from a beer-bottle with a long neck that could be put through a mesh of the wires. They preferred this mode of drinking, and appeared to look upon it as an advanced idea. Elisheba always insisted on being served first, and being a female her wish was complied with. When she had finished, Aaron would climb up by the wires and take his turn. There is a certain sound or word which the chimpanzee always uses to express "good" or "satisfaction," and he made frequent use of it. He would drink a few swallows of the water and then utter the sound, whereupon Elisheba would climb up again and taste it. She seemed to think it was something better than she was drinking, but finding it the same as she had had, she would again give way for him. Every time he would use the sound she would take another taste and turn away, but she never failed to try it if he uttered the sound.

The boy who cared for them on the voyage was disposed to play tricks on them, and one of these ugly pranks was to turn the bottle up so that when they had finished drinking and took their lips away, the water would spill out and run down over them. For a time or two they declined to drink from the bottle while he was holding it, but when he let it go it would hang in such a position that they could not get the water out of it at all. At length Aaron solved the problem by climbing up one side of the cage, and getting on a level with the bottle, reached across the angle formed by the two sides of the cage and drank. In this position it was no matter to him how much the water ran out, it couldn't touch him. Elisheba watched him until she quite grasped the idea, when she climbed up in the same manner and slaked her thirst.

I scolded the boy for serving them with such cruel tricks, but it taught me another lesson of value concerning the mental resources of the chimpanzee, for no philosopher could have found a much better scheme to obviate the trouble than did this cunning little sage in the hour of necessity.

I have never regarded the training of animals as the true measure of their mental powers, but the real test is to reduce the animal to his own resources, and see how he will render himself under conditions that present new problems. Animals may be taught to do many things in a mechanical way, and without any motive that relates to the action; but when they can work out the solution without the aid of man, it is only the faculty of reason that can guide them.

One thing that Aaron could never figure out was what became of the chimpanzee that he saw in a mirror. I have seen him hunt for that mysterious ape for an hour at a time, and he broke a piece off a mirror I had in trying to find it, but he never succeeded.

I have held the glass firmly before him, and he would put his face up close to it, sometimes almost in contact. He would quietly gaze at the image, and then reach his hand around the glass to feel for it. Not finding it, he would peep around the side of it and then look into it again. He would take hold of it and turn it around; lay it on the ground, look at the image again, and put his hand under the edge of it. The look of inquiry in that quaint face was so striking as to make one pity him. But he was hard to discourage, and continued the search whenever he had the mirror.

Elisheba never worried herself much about it. When she saw the image in the glass she seemed to recognise it as one of her kind, but when it would vanish she let it go without trying to find it. In fact, she often turned away from it as though she did not admire it. She rarely ever took hold of the glass, and never felt behind it for the other ape.

Altogether she was an odd specimen of her tribe, eccentric and whimsical beyond anything I have ever known among animals, yet with all her freaks Aaron was fond of her, and she afforded him company; but he was extremely jealous of her, and permitted no stranger to take any liberties with her with impunity. He did not object to them doing so with him, and rarely took offence at any degree of familiarity, for he would make friends with any one who was gentle with him, but he could not tolerate their doing so with her.

She betrayed no sign of affection for him except when some one annoyed or vexed him, but in that event she never failed to take his part against all odds. At such times she would become frantic with rage, and if the cause was prolonged, she would often refuse to eat for hours afterwards.

On the voyage homeward, there was another chimpanzee on board, belonging to a sailor who was bringing him home for sale. He was about two years older than Aaron and fully twice as large. He was tame and gentle, but was kept in a close cage to himself. He saw the others roaming about the deck and tried to make up with them, but they evinced no desire to become intimate with one who was confined in such a manner.

One bright Sunday morning, as we rode the calm waters near the Canary Islands, I induced the sailor to release his prisoner on the main deck with my own, and see how they would act towards each other. He did so, and in a moment the big ape came ambling along the deck towards Aaron and Elisheba, who were sitting on the top of a hatch and absorbed in gnawing some turkey bones.

As the stranger came near he slackened his pace and gazed earnestly at the others. Aaron ceased eating and stared at the visitor with a look of surprise, but Elisheba barely noticed him. He scanned Aaron from head to foot, and Aaron did the same with him. He advanced until his nose almost touched that of Aaron, and in this position the two remained for some seconds, when the big one proceeded to salute Elisheba in the same manner, but she gave him little attention. She continued to gnaw the bone in her hand, and he had no reason to feel flattered at the impression he appeared to have made on her.

Aaron watched him with deep concern, but without uttering a sound.

Turning again to Aaron, he reached out for his turkey bone; but the hospitality of the little host was not equal to the demand, and he drew back with a shrug of his shoulder, holding the bone closer to himself and then resumed eating.

A bone was then given to the visitor by a steward, and he climbed upon the hatch and took a seat on the right of Elisheba, while Aaron was seated to her left. As soon as the big one had taken his seat, Aaron resigned his place and crowded himself in between them. The three sat for a few moments in this order, when the big one got up and deliberately walked around to the other side of Elisheba and sat down again beside her. Again Aaron forced himself in between them.

This act was repeated six or eight times, when Elisheba left the hatch and took a seat on a spar that lay on deck. The big ape immediately moved over and sat down near her; but by the time he was seated Aaron again got in between them, and as he did so he struck his rival a smart blow on the back. They sat in this manner for a minute or so, when Aaron drew back his hand and struck him again. He continued his blows all the while, increasing them in force and frequency, but the other did not resent them. His manner was one of dignified contempt, as if he regarded the inferior strength of his assailant unworthy of his own prowess.

It would be absurd to suppose that he was constrained by any principle of honour, but his demeanour was patronising and forbearing, like that of a considerate man towards a small boy.

One amusing feature of the affair was the half-serious and half-jocular manner of Aaron. He did not turn his face to look at his rival as he struck, and the instant the blow was delivered he withdrew his hand as if to avoid being detected. He gave no sign of anger, but made no effort to conceal his jealousy, and the other seemed to be aware of the cause of his disquietude. The smirk of indifference on the little lover's face belied the state of mind that impelled his action, and it was patent to all who witnessed the tilt that Aaron was jealous of his guest.

From time to time Elisheba would change her seat, when the same scene would ensue.

The whole affair was comical and yet so real, that one could not repress the laughter it evoked. It was the drama of "love's young dream" in real life, in which every man, at some period of his young career, has played each part the same as these two rivals. Every detail of plot and line was the duplicate of a like incident in the experience of boyhood.

AARON AND ELISHEBA

Elisheba did not appear to encourage the suit of this simian beau, but she did not rebuff him as a true and faithful spouse should do, and I never blamed Aaron for not liking it. She had no right to tolerate the attentions of a total stranger; but she was feminine, and perhaps endowed with all the vanity of her sex and fond of adulation.

However, my sympathies for the devoted little Aaron were too strong for me to permit him to be imposed upon by a rival, who was twice as big and three times as strong as he was, so I took him and Elisheba away on the after deck, where they had a good time alone.

Elisheba was never very much devoted to me, but in the early part of her career she began to realise the fact that I was her master and her friend. She had no gratitude in her nature, but she had sense enough to see that all her food and comfort were due to me, and as a matter of policy she became submissive, but never tractable. She was doubtless a plebeian among her own race, and was not capable of being brought up to a high standard of culture. She could not be controlled by kindness alone, for she was by nature sordid and perverse. I was never cruel or severe in dealing with her, but it was necessary to be strict and firm. Her poor health, however, often caused me to indulge her in whims that otherwise would have brought her under a more rigid discipline; and the patient conduct of Aaron appeared to be tempered by the same consideration.

CHAPTER XI
THE DEATH OF AARON AND ELISHEBA

At the end of forty-two long days at sea we arrived at Liverpool. It was near the end of autumn. The weather was cold and foggy. Elisheba was failing in health, as I feared she would do in coming from the warm, humid climate along the equator, and, at the same time, having to undergo a change of food.

On arriving at the end of our long and arduous voyage, I secured quarters for them, and quickly had them stowed away in a warm, sunny cage. Elisheba began to recover from the fatigue and worry of the journey, and for a time was more cheerful than she had been since I had known her. Her appetite returned, the symptoms of fever passed away, and she seemed benefited by the voyage rather than injured. Aaron was in the best of health, and had shown no signs of any evil results from the trip.

On reaching the landing-stage in Liverpool, some friends who met us there expressed a desire to see them, and I opened their cage in the waiting-room for that purpose. When they beheld the throng of huge figures with white faces, long skirts and big coats, they were almost frantic with fear. They had never before seen anything like it, and they crouched back in the corner of the cage, clinging to each other and screaming in terror.

When they saw me standing by them they rushed to me, seized me by the legs, and climbed up to my arms. Finding they were safe here, they stared for a moment, as if amazed at the crowd, and then Elisheba buried her face under my chin, and refused to look at any one. They were both trembling with fright, and I could scarcely get them into their cage again; but after they were installed in their quarters with Dr. Cross, they became reconciled to the sight of strangers in such costumes.

In their own country they had never seen anything like this, for the natives to whom they were accustomed wear no clothing as a rule, except a small piece of cloth tied round the waist, and the few white men they had seen were mostly dressed in white; but here was a great crowd in skirts and overcoats, and I have no doubt that to them it was a startling sight for the first time.

During the first two weeks after arriving at this place, Elisheba improved in health and temper until she was not like the same creature; but about that time she contracted a severe cold. A deep, dry cough, attended by pains in the chest and sides, together with a piping hoarseness, betrayed the nature of her disease, and gave just cause for apprehension.

During frequent paroxysms of coughing she pressed her hands upon her breast or side to arrest the shock, and thus lessen the pain it caused. When quiet, she sat holding her hands on her throat, her head bowed down, and her eyes drooping or closed. Day by day the serpent of disease drew his deadly coils closer and closer about her wasting form, but she bore it with a patience worthy of a human being.

The sympathy and forbearance of Aaron were again called into action, and the demand was not in vain. Hour after hour he sat with her locked in his arms, as he is seen in the portrait given herewith. He was not posing for a picture, nor was he aware how deeply his manners touched the human heart. Even the brawny men who work about the place paused to watch him in his tender offices to her, and his staid keeper was moved to pity by his kindness and his patience.

For days she lingered on the verge of death. She became too feeble to sit up, but as she lay on her bed of straw, he sat by her side, resting his folded arms upon her, and refusing to allow any one to touch her. His look of deep concern showed that he felt the gravity of her case, in a degree that bordered on grief. He was grave and silent, as if he foresaw the sad end that was near at hand. My frequent visits were a source of comfort to him, and he evinced a pleasure in my coming that bespoke his confidence in me and faith in my ability to relieve his suffering companion; but, alas! she was beyond the aid of human skill.

On the morning of her decease, I found him sitting by her as usual. At my approach he quietly rose to his feet, and advanced to the front of the cage. Opening the door, I put my arm in and caressed him. He looked into my face, and then at the prostrate form of his mate. The last dim sparks of life were not yet gone out, as the slight motion of the breast betrayed, but the limbs were cold and limp. While I leaned over to examine more closely, he crouched down by her side and watched with deep concern to see the result. I laid my hand upon her heart to ascertain if the last hope was gone; he looked at me, and then placed his own hand by the side of mine, and held it there as if he knew the purport of the act.

Of course, to him this had no real meaning, but it was an index to the desire which prompted it. He seemed to think that anything that I did would be good for her, and his purpose, doubtless, was to aid me. When I removed my hand, he removed his; when I returned mine, he did the same; and to the last gave evidence of his faith in my friendship and good intentions. His ready approval of anything I did showed that he had a vague idea of my purpose.

At length the breast grew still and the feeble beating of the heart ceased. The lips were parted and the dim eyes were half-way closed, but he sat by as if she were asleep. The sturdy keeper came to remove the body from the cage; but Aaron clung to it, and refused to allow him to touch it. I took the little mourner in my arms, but he watched the keeper jealously, and did not want him to remove or disturb the body. It was laid on a bunch of straw in front of the cage and he was returned to his place, but he clung to me so firmly that it was difficult to release his hold. He cried in a piteous tone, fretted and worried, as if he fully realised the worst. The body was then removed from view, but poor little Aaron was not consoled. How I pitied him! How I wished that he was again in his native land, where he might find friends of his own race!

After this, he grew more attached to me than ever, and when I went to visit him he was happy and cheerful in my presence; but the keeper said that while I was away he was often gloomy and morose. As long as he could see me or hear my voice, he would fret and cry for me to come to him. When I would leave him, he would scream as long as he had any hope of inducing me to return.

A few days after the death of Elisheba, the keeper put a young monkey in the cage with him for company. This gave him some relief from the monotony of his own society, but never quite filled the place of the lost one. With this little friend, however, he amused himself in many ways. He nursed it so zealously and hugged it so tightly that the poor little monkey was often glad to escape from him in order to have a rest. But the task of catching it again afforded him almost as much pleasure as he found in nursing it.

Thus he passed his time for a few weeks, when he was seized by a sudden cold, which in a few days developed into an acute type of pneumonia.

I was in London at the time and was not aware of this, but, feeling anxious about him, I wrote to Dr. Cross, in whose care he was left, and received a note in reply, stating that Aaron was very ill, and not expected to live. I prepared to go to visit him the next day, but just before I left the hotel I received a telegram stating that he was dead.

The news contained in the letter was a greater shock to me than that in the telegram, for which, in part, the former had prepared me; but no one can imagine how deeply these evil tidings affected me. I could not bring myself to a full sense of the fact. I was unwilling to believe that I was thus deprived of my devoted friend. I could not realise that fate would be so cruel to me; but, alas! it was true.

Not being present during his short illness or at the time of his death, I cannot relate any of the scenes attending them; but the kind old keeper who attended him declares that he never became reconciled to the death of Elisheba, and that his loneliness preyed upon him almost as much as the disease.

When I looked upon his cold, lifeless body, I felt that I was indeed bereft of one of the dearest and one of the most loyal pets that any mortal had ever known. His fidelity to me had been shown in a hundred ways, and his affections had never wavered. How could any one requite such integrity with anything unkind?

To those who possess the higher instincts of humanity, it will not be thought absurd in me to confess that the conduct of these creatures awoke in me a feeling more exalted than a mere sense of kindness. It touched some chord of nature that yields a richer tone; but only those who have known such pets as I have known them can feel towards them as I have felt.

I have no desire to bias the calm judgment or bribe the sentiment of him who scorns the love of nature, by clothing these humble creatures in the garb of human dignity; but to him who is not so imbued with self-conceit as to be blind to all evidence and deaf to all reason, it must appear that they are gifted with like faculties and passions to those of man; differing in degree, but not in kind.

Moved by such conviction, who could fail to pity that poor, lone captive, in his iron cell, far from his native land, slowly dying? It may be a mere freak of sentiment that I regret not being with him to soothe and comfort his last hours, but I do regret it deeply. He had the right to expect it of me, as a duty.

Poor little Aaron! In the brief span of half a year he had seen his own mother die at the hands of the cruel hunters; he had been seized and sold into captivity; he had seen the lingering torch of life go out of the frail body of Moses; he had watched the demon of death bind his cold shackles on Elisheba; and now he had, himself, passed through the deep shadows of that ordeal.

What a sad and vast experience for one short year! He had shared with me the toils and dangers of the sea and land over many a weary mile. He seemed to feel that the death of his two friends was a common loss to us; and if there is any one thing which more than another knits the web of sympathy about two alien hearts, it is the experience of a common grief.

Thus ended the career of my *kulu-kamba* friend, the last of my chimpanzee pets. In him were centred many cherished hopes, but they did not perish with him, for I shall some day find another one of his kind in whom I may realise all that I had hoped for in him; but I cannot expect to find a specimen of superior qualities, for he was certainly one of the jolliest and one of the wisest of his race.

However fine and intelligent his successor may be, he can never supplant either Moses or Aaron in my affections: for these two little heroes shared with me so many of the sad vicissitudes of time and fortune that I should be an ingrate to forget them or allow the deeds of others to dim the glory of their memory.

I have all of them preserved, and when I look at them the past comes back to me, and I recall so vividly the scenes in which they played the leading *rôles*—it is like a panorama of their lives.

CHAPTER XII
OTHER CHIMPANZEES

Among the number of chimpanzees that I have seen are some whose actions are worthy of record; but as many of them were the repetition of similar acts of other specimens which are elsewhere described, we shall omit them, and relate only such other acts as may tend to widen the circle of our knowledge, and more fully illustrate the mental range of this interesting tribe of apes.

In passing through the country of the Esyira tribe, I came to a small village where I halted for a rest. On entering the open space between two rows of bamboo huts, I saw a group of native children at the opposite end of the space, and among them a fine big chimpanzee, who was sharing with them in their play.

When they discovered the presence of a white man in the town, they left their sport and came to inspect me. The ape also came, and he showed as much interest in the matter as any one else did. I was seated in a native chair in front of the king's hut, and the people, as usual, stood around me at a respectful distance, looking on as if I had been some wild beast captured in the jungle. The ape was aware that I was not a familiar kind of thing, and he appeared in doubt as to how he should act towards me. He sat down on the ground among the people, and stared at me in surprise, from time to time glancing at those around him as if to ascertain what they thought of me. As they became satisfied with looking, they retired one by one from the scene, until most of them had gone, but the ape remained. He changed his place a few times, but only to get a better view. The people were amused at his manner, but no one molested him.

At length I spoke to him in his own language, using the sound which they use for calling one another. He looked as if he knew what it meant but made no reply. I repeated the sound, when he rose up and stood on his feet as if he intended to come to me. Again I uttered it, and he came a few feet closer, but shied to one side as if to flank my position and get behind me. He stopped again to look, and I repeated the word, in response to which he came up near my right side, and began to examine my clothing. He plucked at my coat-sleeve a few times, then at the leg of my trousers and at the top of my boot. He was getting rather familiar for a stranger, but I felt myself to blame for having given him the license to do so. For a while he continued his investigations, then deliberately put his left hand on my right shoulder, his right foot on my knee, and climbed into my lap. He now began to examine my helmet, ears, nose, chin and mouth. He became a little rough, and I tried to get him down out of my lap, but he was not disposed to go. Finally, I told my boy, who acted as interpreter, to tell the native lads to come and take him away. This amused them very much, for they saw that I was bigger than the ape, and thought I ought therefore to manage him myself. They complied, however, but his apeship declined to go until one of the men of the town interfered and compelled him to do so.

As he got down from my lap, one of the boys bantered him to play. He accepted the challenge, and ran after the lad until they reached the end of the open space between the houses, when the boy fell upon the ground and the ape fell on him. They rolled and wallowed on the ground for a time, when the ape released himself and ran away to the other end of the opening, the boy pursuing him. When they reached the end of the street, they again fell upon each other and another scuffle ensued. It was plain to be seen that the boy could run much faster than the ape, but he did not try to elude him.

The other children crowded around them or followed them, looking on, laughing and shouting in the greatest glee. First one boy and then another took his turn in the play, but the ape did not lose interest in me. He stopped from time to time to take another survey, but did not try again to get upon my lap.

NATIVE VILLAGE AT MOILE—INTERIOR OF NYANZA

After a long time at this sport, the ape quit playing and sat down by the wall of a house, with his back against it; the children tried in vain to induce him to resume, but he firmly declined, and sat there like a tired athlete, picking his teeth with a bamboo splinter, which he had pulled off the side of the house.

His conduct was so much like that of the children with whom he was playing, that one could not have distinguished him from them except by his physique. He enjoyed the game as much as they did, and showed that he knew how to gain or use an advantage over his adversary. In a scuffle he was stronger and more active than the boys, but in the race they were the more fleet. He screamed and yelled with delight, and in every way appeared to enter into the spirit of the fun.

He was about five years old, and his history, as it was given to me, was that he had been captured when quite young in the forest near that place and ever since that time had lived in the village. He had been the constant playmate of the children, ate with them, and slept in the same houses with them. He was perfectly tame and harmless; he knew every one in the village by name, and knew his own name.

The king's son, to whom he belonged, assured me that the ape could talk, and that he himself could understand what he said; but he declined to gratify my request to hear it. However, he called the ape by name, and told him to come to him, which he obeyed. He then gave him a long-necked gourd, and told him to go to the spring and bring some water. The animal hesitated, but on repeating the command two or three times, he reluctantly obeyed. After a few minutes he returned with the gourd about half filled with water. In carrying the vessel he held it by the neck, but this deprived him of the use of one hand. He waddled along on his feet, using the other hand, but now and then would set the gourd on the ground, still holding to it, and using it something after the manner of a short stick. On delivering the gourd of water to his master, he gave evidence of knowing that he had done a clever thing. I expressed a desire to see him fill the gourd at the spring. The water was then emptied out, and the gourd again given to him. On this occasion we followed him to the place where he got the water. On arriving, he leaned over the spring and pressed the gourd into the water, but the mouth of it was turned down so that the water could not flow into it. As he lifted the gourd out, it turned to one side and a small quantity flowed into it. He repeated the act a number of times, and seemed to know how it ought to be done, although he was very awkward in doing it. Whenever the water in the mouth of the gourd would bubble, he would dip it back again and was evidently aware that it was not filled. Finally, raising the vessel, he turned and offered it to his master, who declined to relieve him of it. We turned to go back into the town, and the ape followed us with the gourd, but all the way along continued to mutter a sound of complaint.

He next sent him into the edge of the forest to bring firewood. He was only gone a few minutes when he returned with a small branch of dead wood which he had picked up on the ground. He again sent him, together with three or four children. When he returned on this occasion he had three sticks in his hand. The man explained to me that, when the ape went alone he would never bring but one twig at a time, and this was sometimes not bigger than a lead-pencil; but if the children went with him and brought wood, he would bring as much as he could grasp in one hand. He also told me that the animal would sit down on the ground and lay the sticks across one arm in the same manner as the children did, but invariably dropped them when he would rise up. Then he would seize what he could in one hand, and bring it along. He also said, that in carrying a single stick the ape always used the hand in which he held it; but if he had three or four pieces that he always curved his arm inwards, holding the wood against his side, and hobbled along with his feet and the other hand.

The next thing with which he entertained me was sending the ape to call some one in the village. He first sent him to bring a certain one of the man's wives. She was several doors away from where we sat. The ape went to one house, sat down at the door for a moment, looking inside, and then moved slowly along to the next, which he entered. Within a minute he appeared at the door holding the cloth that the woman wore tied around her, and in this manner led her to his master. He next sent him to bring a certain boy, which he did in a similar manner, except that the boy had on no clothing of any kind, and the ape held him by the leg.

During all these feats the man talked to him, as far as I could tell, in the native language only, though he declared to me that some of the words that he had used were those of the ape's own speech. However, he said that many words that the ape knew were of the native speech, and that the ape had no such words in his language. One thing that especially impressed me was a sound which I have elsewhere described as meaning "good" or "satisfaction," which this man said was the word which these apes use to mean "mother." My own servant had told me the same thing before, but I am still of the opinion that they are mistaken in the meaning of the sound, although it is almost exactly the same as the word for mother in the native speech. The difference being in the vowel element only, and it is possible, I grant, that the word may have both meanings. A little later one of the women came to the door of a house and said, in the native language, that something was ready to eat, whereupon the children and the ape at once started. In the meantime she set an earthen pot, containing boiled plantains, in front of the house, from which all the children and the ape alike helped themselves. In brief, the ape was a part of the family, and was so regarded by all in the town.

I do not know to what extent they may have played upon my credulity, but, so far as I could discern, their statements concerning the animal were verified.

I proposed to buy the ape, but the price asked was nearly twice that of a slave, and I could have bought any child in the town at a smaller cost. I have never seen any other chimpanzee that I so much coveted. When standing in an upright position, he was quite four feet in height, strongly built, and well-proportioned. He was in a fine, healthy condition, and in the very prime of his life. He was not handsome in the face, but his coat of hair was of good colour and texture. He was of the common variety, but a fine specimen.

Mr. Otto Handmann, formerly the German consul at Gaboon, had a very fair specimen of this same species of chimpanzee. He was a rough, burly creature, but was well-disposed and had in his face a look of wisdom that was almost comical. He had been for some months a captive in a native town, during which time he had become quite tame and docile. By nature he was not humorous, but appeared to acquire a sense of fun as he grew older and became more familiar with the manners of men.

On my return from the interior, I was invited by the consul to take breakfast with himself and a few friends; but owing to a prior engagement I was not able to be present. It was proposed by some one of the guests present that my vacant seat at the table should be filled by the chimpanzee. He was brought into the room and permitted to occupy the seat. He behaved himself with becoming gravity, and was not abashed in the presence of so many guests. He was served with such things as were best suited to his liking, and his demeanour was such as to amuse all present. On proposing a toast, all the guests beat with their hands upon the table, and in this the chimpanzee joined with apparent pleasure. After a few rounds of this kind, one of the guests, occupying the seat next to him, failed to respond with the usual beating; the chimpanzee observed the fact, turned upon the guest, and began to claw, scream, and pound him on the back and arm until the gentleman proceeded to beat, whereupon the ape resumed his place and joined in the applause. On this occasion he acquitted himself with credit, but an hour later he had fallen into disgrace by drinking beer until he was actually drunk, when he awkwardly climbed off the chair, crawled under the table, and went to sleep.

One of the clerks in the employ of the consul also had a fair specimen of this same species. It was a female, perhaps two years younger than the one just described, but equally addicted to the habit of drinking beer. It is the custom among people on the coast to offer to a guest something to drink, and on these occasions this young lady ape always expected to partake with others. If she was overlooked in pouring out beer for others, she always set up a complaint until she got her glass. If it was not given to her, she would go from one to another, holding out her hand and begging for a drink. If she failed to secure it, she would watch her opportunity, and while the guest was not looking, would stealthily reach up, take his glass off the table, drink the contents, and return the glass to its place. She would do this with each one in turn, until she had taken the last glass; but if a glass was given to her at the same time that the others were served, she was content with it and made no attempt to steal that of another.

In this act she evinced a skill and caution worthy of a confirmed thief; she would secrete herself under the table or behind a chair, and watch her chance. She made no attempt to steal the glass while it was being watched, but the instant she discovered that she was not observed, or thought she was not, the theft was committed.

Her master frequently gave her a glass and bottle of beer to help herself. She could pour the beer out with dexterity. She often spilt a portion of it, and sometimes filled the glass too full, but always set the bottle right end up, lifted the glass with both hands, drained it, and refilled it as long as there was any in the bottle. She could also drink from the bottle, and would resort to this if no glass was given her. She knew an empty bottle from one that contained beer.

This ape was very much attached to her master, would follow him, and cry after him like a child. She was affectionate to him, but had been so much annoyed by strangers that her temper was spoiled and she was irritable.

I may remark here, that I have known at least five or six chimpanzees that were fond of beer, and would drink it until they were drunk whenever they could get it. I have never seen one, that I am aware of, that would drink spirits.

Arriving on the south side of Lake Izanga, I found a young chimpanzee at the house of a white trader. It was tied to a post in the yard, where it was annoyed by the natives who came to the place to trade. On approaching it for the first time, I spoke to it in its own language, using the word for food. It recognised the sound at once and responded to it. As I came nearer, it advanced as far towards me as the string with which it was tied would allow. Standing erect and holding out its hands, it repeated the sound two or three times. I gave it some dried fish which it ate with relish, and we at once became friends. Its master permitted me to release it on the condition that I should not allow it to escape. I did so, and took the little captive in my arms. It put its arms around my neck as if I had been the only friend it had on earth. It clung to me, and would not consent for me to leave it. I could but pity the poor, neglected creature. There it was, tied in the hot sun, hungry, lonely, and exposed to the tortures of every heartless native that chose to tease it. When it was not in my arms, it followed me around and would not leave me for a moment. Its master cared but little for it, and left it to the charge of his boy, who, like all other natives, had no thought or concern for the comfort of any creature but himself. I tried to purchase it, but the price was too much, and after two days our friendship was broken for ever. But I was glad to learn, soon after this, that another trader secretly released it, and let it escape into the forest. The man who did this told me himself that he did it as an act of mercy. I often recall this little prisoner to mind, and always feel a sense of gladness at knowing that he was set at liberty by a humane friend. Whatever may have been his fate in the forest, it could have been no worse than to be confined, starved, and tormented as he was, while in captivity.

Another small specimen, which I saw at Gaboon, was not of much value except from one fact, and that was, it was broken out with an eruptive disease prevalent among the natives. It is called crawcraw or kra-kra. It is said to originate from the water, either by external or internal use of it. This animal was infected in the same way and on the same parts of the body as men are affected by the same disease, and is another instance of their being subject to the same maladies as those of man. The specimen itself also exemplified the difference in intellect among these animals, for this one had in its face the look of mental weakness, and every act confirmed the fact. It was silent, inactive and obtuse.

During my residence in the cage I did not see so many chimpanzees as I saw of gorillas, but from those I did see it was an easy matter to determine that they were much less shy and timid than the gorilla.

On one occasion I heard one in the bush not far away from the cage. I called him with the usual sound and he answered, but did not come to the cage. It is probable that he could see it, and was afraid of it. I tried to induce Moses to call him, and he did once utter the sound, but he appeared to regret having made the attempt. I called again and he answered, and from the manner in which Moses behaved it was evident that he understood it. He would not attempt the call again, but clung to my neck with his face buried under my chin. It was probably jealousy that caused him to refuse, because he did not want the other to share my attentions. I gave the food sound, but I could not induce the visitor to come nearer. I failed to get a view of him so as to tell how large he was, but from his voice he must have been about grown. Whether he was quite alone or not I was not able to tell, but only the one voice could be heard.

Another time, while sitting quite alone, a young chimpanzee, perhaps five or six years old, appeared at the edge of a small opening of the bush. He plucked a bud or leaf from a small plant. He raised it to his nose and smelt it. He picked three or four buds of different kinds, one or two of which he put in his mouth. He turned aside the dead leaves that were lying on the ground as if he expected to find something under them. I spoke to him, using the call sound; he instantly turned his eyes towards me, but made no reply. I uttered the food sound and he replied, but stood where he was. He betrayed no sign of fear, and little of surprise. He surveyed the cage and myself, and I repeated the sound two or three times. He refused to approach any nearer. He turned his head from side to side for a moment as if in doubt which way to go; then turned aside and disappeared in the bush. He did not run or start away as if in great fear, but by the sound of the shaking bushes it could be told that he increased his speed after he once disappeared from view.

One day I had been for a stroll with Moses and the boy. As we returned to the cage we saw a chimpanzee about half-grown; he was crossing the rugged little path about thirty yards away from us. He paused for a moment to look at us, and we stopped. I tried to induce Moses to call out to him, but he declined to do so. As the stranger turned aside I called to him myself, but he neither stopped nor answered. This one appeared to be quite brown, but the boy assured me his hair was jet black, but his skin being light gave him this colour. To satisfy myself, I had Moses placed in the same place and position, and looking at him from the same distance I was convinced that the boy was right.

One morning, as I started with Moses for a walk, I had only gone some forty yards away from the cage when he made a sound of warning. I instantly looked up, when I saw a large chimpanzee standing in the bush not more than twenty yards away. I paused to look at him. He stood for a moment, looking straight at us. I spoke to him, but he made no reply; he moved off almost parallel to the little path which we were in, and I returned towards the cage. He did not come any nearer to us, but kept his course almost parallel with ours. He turned his head from time to time to look, but gave no sign of attack. I called to him several times, but he made no answer. When I reached a place in front of the cage I called again, and after the lapse of a few seconds he stopped. By this time he was concealed from view. He only halted for a moment, changed his course and resumed his journey. This was the largest one I saw in the forest.

At another time, while sitting in the cage, I heard the sound of something making its way through the bush not more than twenty yards away; presently it passed in view. As it crossed the path near by, I called three or four times, but it neither stopped nor answered. As well as I could tell, it appeared to be a female and quite grown.

I may take occasion to remark that while the chimpanzee is mostly found in large family groups, as I have reason to believe from native accounts of them, and from what has been told me by white men, I have never been able to see a family of them together, but each of these that I have mentioned, so far as I could tell, was quite alone. Whether the others were scattered through the forest in like manner, hunting for food, and all came together after this or not, I can only say that every chimpanzee that I saw was alone at the time.

Another thing worthy of mention is the fact that both these apes live in the same forest, and twice on the same day I have seen both kinds. This is contrary to the common idea that they do not inhabit the same jungle. It appears that where there is a great number of the one there are but few of the other. The natives say that in combat between the chimpanzee and gorilla, the former is always victor, on which account the latter is afraid of him. I believe this to be true, because the chimpanzee, although not so strong, is more active and more intelligent than the gorilla.

The chimpanzee will not approach or attack man if he can avoid it, but he does not shrink from him as the gorilla does. One instance that will illustrate this phase of his character I shall relate. On one occasion recently, while I was on the coast, a native boy started across a small plain near the trading station. Along with him was a dog that belonged to the white trader at the place. The dog was in advance of the boy, and as the latter emerged from a small clump of the bush he heard the dog bark in a playful manner, and discovered him not more than thirty yards away, prancing, jumping, and barking in a jolly way with a chimpanzee which appeared to be five or six years old. The ape was standing in the path along which the boy was proceeding. He was slapping at the dog with his hands, and did not seem to relish the sport, yet he was not resenting it in anger. The dog thought the ape was playing with him, and he was taking the whole thing in fun. The boy looked at them for a few moments and retreated. As soon as he disappeared the dog desisted and followed him to the house. The boy was afraid of the ape, and made no attempt to capture him. The latter was taken by surprise by the dog and boy, and thus had no time to escape. He did not strike to harm the dog, but only to ward him off. The dog made no attempt to bite him, but when he would jump up against him he would knock the ape out of balance, and this annoyed him. He didn't seem to understand just what the dog meant.

I shall not describe those so well known in captivity, only to mention some of them. The largest specimen of the chimpanzee that I have ever seen was Chico, who belonged to Mr. James A. Bailey, of New York. He was as large perhaps as these apes ever become, although he was less than ten years old when he died.

Perhaps the most valuable specimen for scientific use that has ever been in captivity is Johanna, who belongs to the same gentleman. The history that is given of her, however, is hardly to be taken in full faith. Her age cannot be determined with certainty, but it is said that she is about thirteen years old. I have reason to doubt that, although I cannot positively deny it. Whatever may be her exact age, it is certain that she has now reached a complete adult state. She has grown to be quite as large as Chico was at the time of his death. She is not of amiable temper, but is much less vicious than he was. She has some of the marks of a kulu kamba.

In order to justify my doubts upon the subject of her age, I may state that Chico was only ten years of age when he died, but had reached the adult period; and as males do not reach that state sooner than the females of any genus of the primates, it is not probable that he was mature at ten, while she was not so until twelve. In the next place, her captors claim to have seen her within a few hours after her birth, and that they watched her and her mother from time to time until she was one year old, when they killed the mother and captured the babe. The claim is absurd. These apes are nomadic in habit, and are rarely ever seen in the same place. They claim that she was born on January 19, but from what I know of these apes that is not their season of bearing, and I doubt if any of them were ever born during that month. Again, it is claimed that she was captured by Portuguese explorers in the Congo, but the Portuguese do not possess any territory along that river in which these apes are ever found. They claim the territory around Kabenda, which would indicate that she came from the Loango Valley instead of the Congo, but the cupidity of the average Portuguese would never allow anything to go at liberty for a year if it could be sold before that time.

Johanna is accredited with a great deal of intelligence; but I do not regard her as being above the average of her race. Since the death of her companion, Chico, she has received the sole attention of her keeper, and since that time has been taught a few things which are neither marvellous nor difficult. In point of intellect she cannot be regarded as an extraordinary specimen of her tribe. I do not mean to detract from her reputation, but I have failed to discover in her any high order of mental qualities.

The reason why Johanna may be regarded as the most valuable specimen for study is the fact that she is the only female of her race that has ever reached the state of puberty. She has done so, and this fact enables us to determine certain things which have never heretofore been known. This affords the Zoologists an opportunity for the study of her sexual development which may not again present itself in many years to come. From this important point of view she presents the student with many new problems in that branch of science.

I have elsewhere stated as my opinion that the female chimpanzee reaches the age of puberty at seven to nine years, and I have many reasons which I will not here recount, that cause me to adhere to that belief. But the uncertainty of the age of this ape does not destroy her value as a subject of scientific study.

The most sagacious specimen of the race that I have been brought in contact with is Consul II., who is now an inmate of the Bellvue Garden of Manchester, England. He has not been educated to perform mere tricks to gratify the visitor in the way that animals are usually trained, but most of the feats that he performs are prompted by his own desire and for his own pleasure.

CONSUL II. RIDING A TRICYCLE

There is a vast difference in the motives that prompt animals in the execution of these feats. I have elsewhere mentioned the fact that animals that are caused to act from fear do so mechanically, and it is not a true index to their intellect. While Consul and a few other apes that I have seen do many things by imitation they do not do so from coercion. They seem to understand the purpose and foresee the results, and these impel them to act.

Some of the feats performed by this ape I have never seen attempted by any other. One accomplishment is riding a tricycle. He knows the machine by the name of "bike," although it is not really a bicycle. He can adjust it and mount it with the skill of an acrobat. The ease and grace with which he rides are sufficient to provoke the envy of any boy in England. He propels it with great skill and steers it with the accuracy of an expert. He guides it around angles and obstacles in the way with absolute precision.

Consul is allowed to go at liberty a great deal of his time, which is the proper way to treat these apes in captivity. He rides the wheel for his own diversion. He does not do it to gratify strangers or to "show off."

Another accomplishment which he has, is that of smoking a pipe, cigar, or cigarette. It may not be commended from a moral standpoint, but the act appears to afford him quite as much pleasure as it does the average boy when he first acquires it, and he has also formed the habit of spitting as he smokes, but he has the good manners not to spit on the floor. When Consul has his pipe lighted he usually sits on the floor to enjoy it, and he spreads a sheet of paper down before him to spit on. When he has finished smoking he rolls up the paper and throws it into some corner out of the way. When playing about the grounds he often finds a cigar stub. He knows what it is, picks it up, puts it into his mouth and at once goes to his keeper for a light. He will not attempt to light his pipe or cigar, because he is afraid of burning his fingers; but he will light a match and hand it to his keeper to hold while lighting the pipe. He sometimes takes a piece of paper, lights it in the fire and hands it to some one else to light his pipe for him. He is afraid of the fire, and will not hold the paper while it is burning. If any one hesitates to take it from him, he throws it at them and gets out of the way. He is not so fond of cigarettes, because he gets the tobacco in his mouth, and he does not like the taste of it.

When Consul is furnished with a piece of chalk, he begins to draw some huge figure on the wall or floor. He never attempts to make a small design with chalk, but if given a pencil and paper he executes some peculiar figure of smaller design. Those made with the chalk or pencil are usually round or oval in shape, but if given a pen and ink he at once begins to make a series of small figures containing many acute angles. Whether these results are from design or accident I cannot say, but he appears to have a well-defined idea as to the use of the instrument, but whether he can distinguish between writing and drawing I am unable to say.

The only abstract thing that his keeper has tried to teach him is to select the letters of the alphabet. He has learned to distinguish the first three. These are made upon the faces of cubical blocks of wood: each block contains one letter on each of its faces. He selects the letter asked for with very few mistakes, and this appears to be from indifference more than from ignorance.

Consul is very fond of play, and makes friends with some strangers on sight, but to others he takes an aversion without any apparent cause, and while he is not disposed to be vicious when not annoyed, he resents with anger the approaches of certain persons. He is the only one I have seen that can use a knife and fork with very much skill, but he cuts up his food with almost as much ease as a boy of the same age would do, and uses his fork in eating. He has been taught to do this until he rarely uses his fingers in the act. He is fond of coffee and beer, but does not care for spirits.

There is nothing that so much delights Consul as to get into the large cage of monkeys and baboons kept in the garden. Most of them are afraid of him. But one large Guinea baboon is not, and on every occasion he shows his dislike for the ape. The latter, however, takes many chances in teasing him, but always manages to evade his attack. He displays much skill and a great degree of caution in playing these pranks upon the baboon when at close range. Upon the approach of the ape the other animals in the cage all seek some refuge, and he finds great diversion in stealing up to their place of concealment to frighten them. Consul is very strong, and can lift objects of surprising weight. It is awkward for him to stand in an upright position, but he does so with more ease than any other chimpanzee that I have ever seen. If any one will take hold of his hand he will stroll with him for a long time without apparent fatigue.

Owing to the sudden changes of temperature in that part of England, he is provided with a coat, which he is often required to wear when going out of doors. He does not like to be hampered with such garments, and if for a moment he is not watched, he removes it, and sometimes hides it to keep from wearing it. He is also provided with trousers, which he dislikes more if possible than his coat; but above all other articles of wearing apparel he dislikes shoes. His keeper often puts them on him, but whenever he gets out of sight he unties and removes them. He cannot tie the laces, but can untie them in an instant.

CONSUL II. IN FULL DRESS

He does not evince so much aversion to a hat or cap, and will sometimes put one on without being told; but he has a perfect mania for a silk hat, and if allowed to do so he would demolish that of every stranger who comes to the garden. He has a decided vein of humour and a love of approbation. When he does anything that is funny or clever, he is perfectly aware of the fact; and when by any act he evokes a laugh from any one he is happy, and recognises the approval by a broad chimpanzee grin.

In the corner of the monkey-house is a room set apart for the keeper, and in this room supplies of food for the inmates are kept. In a small cupboard in one corner is kept a supply of bananas and other fruits. Consul knows this and has tried many times to burglarise it. On one occasion he secured a large screw-driver and attempted to prise open the door. He found the resistance to be greatest at the place where the door locked, and at this point he forced the instrument in the crevice and broke off a piece of the wood about an inch wide from the edge of the door. At this juncture he was discovered and reproved for his conduct, but he never fails to stick his fingers in this crack and try to open the door. He has not been able to unlock it when the key is given him, although he knows the use of it, and has often tried, but his keeper has never imparted the secret to him, and his method of using the key has been to prise with it, or pull it instead of turning it after putting it in the keyhole.

The young keeper, Mr. Webb, deserves great credit for his untiring attention to this valuable young ape, and the results of his zeal are worthy of the recognition of every man who is interested in the study of animals.

Another specimen that may be regarded as an intermediate type was recently kept in Belle Vue Gardens at Manchester. He was playful and full of mischief. He had been taught to use a stick or broom to fight with, and with such a weapon in his hand would run all over the building, hunting some one to fight. He did not appear to be serious in his assault, but treated it as fun. It was a bad thing to teach an ape, because they grow pugnacious as they grow older, and all animals kept closely confined acquire a bad temper.

In an adjoining cage was kept a young orang, and the two ate at the same table. The chimpanzee appeared to entertain a species of contempt for the orang. The keeper had taught him to pass the bread to his neighbour, and he obeyed this with such reluctance that his manner betrayed more disgust than kindness. A few small pieces of bread were placed on a tin plate, and the kulu was required to lift the plate in his hand, and offer it to the orang before he himself was allowed to eat. He would lift the plate a few inches above the table, and hold it before the orang's face; when the latter had taken a piece of the bread, the chimpanzee withdrew the plate, held it for a moment, and dropped it. Meanwhile he kept his eyes fixed on the orang. The manner in which he dropped the plate looked as if he did so in contempt. When the meal was finished, the kulu would drink his milk from a cup, wipe his mouth with the serviette, and then get down from the table. The orang would slowly climb down, and go back to his cage. We shall not describe the details of their home-life, but they were two jolly young bachelors, one of which was as stupid as the other was bright.

The specimens that were kept in the Gardens in New York were very fine. One of them was mentally equal to any other specimen hitherto in captivity. There were two kept in the Cincinnati Gardens which were also very fine. There have never been but nine of these apes brought to America so far as I am aware, but six of these lived longer and four of them grew to be larger than any other specimens of this race have ever done in captivity. For some reason they never survive long in England, or other parts of Europe. This is probably due to some condition of the atmosphere. It cannot be from a difference of treatment.

I have seen a large number of chimpanzees, but most of them were in captivity, yet I have seen enough of them in a wild state to gain some idea of their habits and manner, but those described will be sufficient to show the mental character of the genus.

CHAPTER XIII
OTHER KULU-KAMBAS

Whether the kulu-kamba is a distinct species of ape, or only a well-marked variety of the chimpanzee, he is by far the finest representative of his genus. Among those that I have seen are some very good specimens, and the clever things that I have witnessed them do are sufficient to stamp them as the highest type of all apes.

On board a small river steamer that plies the Ogowe, was a young female kulu that belonged to the captain. Her face was not by any means handsome, and her complexion was the darkest of any kulu I have ever seen. It was almost a coffee-colour. There were two or three spots much darker in shade, but not well defined in outline. The dark spots looked as if they had been artificially put on the face. The colour was not solid, but looked as if dry burnt umber had been rubbed or sprinkled over a surface of lighter brown. Although she was young (perhaps not more than two years old), her face looked almost like that of a woman of forty. Her short, flat nose, big, flexible lips, protruding jaws and prominent arches over the eyes, with a low receding forehead, conspired to make her look like a certain type of human being one frequently sees. This gave her what is known as a dish-face, or a concave profile. She had a habit of compressing her nose by contracting the muscles of the face; curling her lips as if in scorn, and at the same time glancing at those around her as if to express the most profound contempt.

Whatever may have been the sentiment in her mind, her face was a picture of disdain, and the circumstances under which she made use of these grimaces, certainly pointed to the fact that she felt just like she looked. At other times her visage would be covered with a perfect smile. It was something more than a grin, and the fact that it was used only at a time when she was pleased or diverted, showed that the emotion which gave rise to it was perfectly in keeping with the face itself. In repose her face was neither pretty nor ugly. It did not strongly depict a high mental status, nor yet portray the instincts of a brute; but her countenance was as safe an index to the mind as that of the human being. This is true of the chimpanzee more perhaps than of any other ape. The gorilla doubtless feels the sense of pleasure, but his face does not yield to the emotion, while the opposite passions are expressed with great intensity, and with the common chimpanzee it is the same way, but not to the same extent.

The kulu in question was more of a coquette than she was of a shrew. She plainly showed that she was fond of flattery. Not perhaps in the same sense that a human being is, but she was certainly conscious of approbation and fond of applause. When she accomplished anything difficult, she seemed aware of it; and when she succeeded in doing a thing which she was not allowed to do, she never failed to express herself in the manner described above. She always appeared to be perfectly conscious of being observed by others, but she was defiant and composed. There was nothing known in the catalogue of mischief that she was not ready to tackle at any moment and take her chances on the result. From the stoke-hole to the funnel, from the jack-staff to the rudder, she explored that boat.

To keep her out of mischief, she was tied on the saloon deck with a long line, but no one aboard the vessel was able to tie a knot in the line which she could not untie with dexterity and ease. Her master, who was a sailor and an expert in the art of tying knots, exhausted his efforts in trying to make one that would defy her skill.

On one occasion I was aboard the little steamer when the culprit was brought up from the main deck where she had been in some mischief, and tied to one of the rails along the side of the boat. The question of tying her was discussed, and at length a new plan was devised. In the act of untying a knot she always began with the part of the knot that was nearest to her. It was now agreed to tie the line around one of the rails on the side of the deck, about half-way between the two stanchions that supported it, then to carry the loose ends of the line to the stanchion and make it fast in the angle of it and the rail. This was done. As soon as she was left alone she began to examine the knots; but she made no attempt at first to untie them except to feel them as if to see how firmly they were made. She then climbed up on the iron rail around which the middle of the line was tied, and slackened the knot. She pulled first at one strand and then at the other, but one end was tied to the stanchion and the other to her neck, and she could find no loose end to draw through. First one way and then the other she drew this noose. She saw that in some way it was connected with the stanchion. She drew the noose along the rail until it was near the post; she climbed down upon the deck, then around the post and back again; she climbed up over the rails and down on the outside, and again carefully examined the knot; she climbed back, then through between the rails and back, then under the rails and back, but she could find no way to get this first knot out of the line. For a moment she sat down on the deck, and viewed the situation with evident concern. She slowly rose to her feet and again examined it; she moved the noose back to its place in the middle of the rail, climbed up by it, and again drew it out as far as the strands would allow. Again she closed it; she took one strand in her hand and traced it from the loop to the stanchion, then she took the other end in the same manner and traced it from the loop to her neck. She looked at the loop and then slowly drew it out as far as it would come. She sat for a while holding it in one hand, and with the other moved each strand of the knot. She was in a deep study, and did not even deign a glance at those who were watching her. At length she took the loop in both hands, deliberately put it over her head and crawled through it. The line thus released dropped to the deck; she quickly descended, took hold of it near her neck, and found that it was untied; she gathered it up as she advanced towards the other end that was tied to the post, and at once began to loosen the knots about it. In a minute more the last knot was released, when she gathered the whole line into a bundle, looked at those around her with that look of contempt which we have described, and departed at once in search of other mischief. The air of triumph and contempt was enough to convince any one of her opinion of what she had done.

If this feat was the result of instinct, the lexicons must find another definition for that word. There were six white men who witnessed the act, and the verdict of all was that she had solved a problem which few children of her own age could have done. Every movement was controlled by reason. The tracing out of cause and effect was too evident for any one to doubt.

NATIVE VILLAGE AT GLASS GABOON

Almost any animal can be taught to perform certain feats, but that does not show the innate capacity. The only true measure of the faculty of reason is to reduce the actor to his own resources, and see how he will render himself under some new condition, otherwise the act will be, at least in part, mechanical or imitative. In all my efforts to study the mental calibre of animals I have confined them strictly to their own judgment, and left them to work out the problem alone. By this means only can we estimate to what extent they apply the faculty of reason. No one doubts that all animals have minds, which are receptive in some degree. But it has often been said that they are devoid of reason, and controlled alone by some vague attribute called instinct. Such is not the case. It is the same faculty of the mind that men employ to solve the problems that arise in every sphere of life. It is the one which sages and philosophers have used in every phase of science. It differs in degree, but not in kind.

This kulu-kamba knew the use of a corkscrew. This she had acquired from seeing it applied by men. While she could not use it herself with success, she often tried and never applied it to the wrong purpose.

She would take the deck broom and scrub the deck, unless there was water on it, in which event she always left the job. She did not seem to know the purpose of sweeping the deck, and never swept the dirt before the broom. This was doubtless imitative. She only grasped the idea that a broom was used to scrub the deck, but she failed to observe the effect produced. However, it cannot be said with certainty to what extent she was aware of the effect, but it is inferred from the fact that she did not try to remove the dirt.

She knew what coal was intended for, and often climbed into the bunker and threw it down by the furnace door. The furnace door and steam gauge were two things that escaped her busy fingers. I do not know how she learned the danger of them, but she never touched them. She had to be watched to keep her from seizing the machinery. For this she seemed to have a strong desire, but did not know the danger she incurred.

I was aboard a ship when a trader brought off from the beach a young kulu to be sent to England. The little captive sat upright on the deck and seemed aware that he was being sent away. At any rate his face wore a look of deep concern as if he had no friend to whom he could appeal. On approaching him I spoke to him, using his own word for food. He looked up and promptly answered it. He looked as if in doubt as to whether I was a big ape or something else. I repeated the sound, and he repeated the answer and came towards me. As he approached me I again gave the sound. He came up and sat by my feet for a moment, looking into my face. I uttered the sound again, when he took hold of my leg and began to climb up as if it had been a tree. He climbed up to my neck and began to play with my lips, nose and ears. We at once became friends, and I tried to buy him, but the price asked was more than I desired to pay. I regretted to part with him, but he was taken back to the beach, and I never saw him again.

On another occasion one was brought aboard, and after speaking to him I gave him an orange; he began to eat it and at the same time caught hold of the leg of my trousers as if he did not wish me to leave him. I petted and caressed him for a moment and turned away, but he held on to me. He waddled about over the deck, holding on to my clothes, and would not release me. He was afraid of his master and the native boy who had him in charge. He was a timid creature, but was quite intelligent, and I felt sorry for him because he seemed to realise his situation.

On the same voyage I saw one in the hands of a German trader. It was a young male, about one year old. He promptly answered the food sound, and I called him to come to me; but this he neither answered nor complied with. He looked at me as if to ask where I had learned his language. I repeated the sound several times, but elicited no answer. I have elsewhere called attention to the fact that these apes do not answer the call when they can see the one who makes it, and they do not always comply with it. In this respect they behave very much the same as young children, and it may be remarked that one difficulty in all apes is to secure fixed attention. This is exactly the same with young children. Even when they clearly understand, sometimes they betray no sign of having heard it. At other times they show that they both hear and understand, but do not comply.

Another specimen that was brought aboard a ship when I was present was a young male, something less than two years old. He was sullen and morose. He did not resent my approaches, but he did not encourage them. I first spoke to him with the food sound, but he gave no heed. I retired a little distance from him and called him, but he paid no attention. I then used the sound of warning; he raised his head, and looked in the direction from which the sound came. I repeated it, and he looked at me for a moment and turned his head away. I repeated it again. He looked at me, then looked around as if to see what it meant, and again resumed his attitude of repose.

On my last voyage to the coast I saw a very good specimen in the Congo. It was a female, a little more than two years old. She was also of a dark complexion, but quite intelligent. She had been captured north of there, and within the limits elsewhere described. At the time I saw her she was ill and under treatment, but her master, the British consul, told me that when she was well she was bright and sociable. I made no attempt to talk with her, except some time after, having left her, I gave the call sound, which she answered by looking around the corner of the house. I do not know whether she would have come or not, as she was tied and could not have done so had she desired to.

I have seen a few other specimens of this ape, and most of them appear to be of a somewhat higher order than the ordinary chimpanzee, but there is among them a wide range of intelligence. It would be a risk to say whether the lowest specimen of kulu is higher or lower than the highest specimen of the common chimpanzee or not, but taken as a whole they are much superior. I shall not describe at length the specimens which have been known in captivity, since most of them have been amply described by others; but it is not out of place to mention some of them.

If proper conditions were afforded to keep a pair of kulus in training for some years, it is difficult to say what they might not be taught. They are not only apt in learning what they are taught, but they are well-disposed, and can apply their accomplishment to some useful end. We cannot say to what extent they may be able to apply what they learn from man, because the necessity of doing so is removed by the attention given them.

CHAPTER XIV
GORILLAS

In the order of nature the gorilla occupies the second place below man. His habitat is in the lowlands of West Tropical Africa, and is confined to very narrow limits. The vague line which bounds his realm cannot be defined with absolute precision, but those generally given in books that treat of him are not correct. If he ever occupied any part of the coast north of the equator, he has long since become extinct in that part, but there is nothing to show that he ever did exist there. So far as I have been able to trace the lines that prescribe his native haunts, he appears to be confined to the low, delta country, lying between the Equator and Loango along the coast, and reaching eastward to the interior, an average distance of about one hundred miles. The eastern boundary is very irregular. To be more exact, the extreme limit on the north side would be the Gaboon River to its head-waters, thence southward to the Ogowe River to the mouth of the Nguni River; up that river twenty or thirty miles, thence a zigzag line along the western base of the dividing lands between the Congo basin and the Atlantic watershed, to the head-waters of the Chi Loango River, and with that to the coast. Beyond these lines I have never been able to find any trace of him, and along this boundary only now and then are they found. I have seen two adult and two infant skulls of the gorilla that were brought by Mr. Wm. S. Cherry, from the Kisango Valley, which lies north of the middle Congo in the interior. The skulls are the only evidence I have ever found of this ape existing so far eastward, but they were said to have come from that part of the valley lying directly under the equator. Mr. Cherry did not collect them himself, but secured them from natives, and does not claim to have seen any of these apes alive.

There appear to be three centres of population: the first is in the basin of Izanga Lake; the second in the basin of Lake Fernan Vaz; and the third in the basin of the lake behind Sette Kama. They are rarely ever found in high or hilly districts, but appear to inhabit the hummock lands, which are only elevated a few feet above tide-level. This is singular, from the fact that the ape has a morbid dislike for deep water, and I think it doubtful if he can swim, although he has one peculiar character that belongs to aquatic animals, which is a kind of web between the digits, but its purpose cannot be to aid in swimming. I have been told that the gorilla can swim, and it may be true; but I have never observed anything in his habits to confirm this, while I have noted many facts that controvert it.

I know of no valid reason why he should be confined so strictly within the limits mentioned, unless it be from a condition of climate which seems peculiar to this district. South of it the climate along the coast is much cooler, and the country back of it is hilly and barren; north of the Equator is a land of perpetual rain, while to the eastward, it is mountainous. Within this district the rainy and dry seasons are more fixed and uniform.

The gorilla appears to be an indigenous product which does not bear transplanting; he thrives only in a low, hot and humid region, infested by malaria, miasma and fevers. It is doubtful if he can long survive in a pure atmosphere.

The only single specimen that I have ever heard of north of the equator, was one on the south side of the Komo River, which is the north branch of the Gaboon. The point at which I heard of him was within a few miles of the equator. I also heard of five having been seen a few miles south-west from Njole, which is located on the Equator on the south side of the Ogowe, a little way east of the Nguni, and they were said to be the first ever seen in that part within the memory of man.

NATIVES SKINNING A GORILLA

As to their being found between Gaboon and Cameroon, I can find no trace along the coast of one ever having been seen in that part. Certain writers have mentioned the fact that in 1851 and 1852 they came in great numbers from the interior to the coast. From such a statement it might be inferred that they were seen in herds or armies together, while the truth was that in those years a few more gorillas appeared to be in the jungle than was usual, but they were not north of the Gaboon River. They were in the Ogowe delta about 1° south latitude; but no one ever supposed that they came from the Crystal Mountains or any other mountains. At that time neither traders nor missionaries had ascended the Gaboon River above Parrot Island (which is less than twenty miles from the mouth), except to make a flying trip by canoe, and nothing was known of that part except what was learned from the natives, and that was very little. During my first voyage I went up that river as far as Nenge Nenge, about seventy-five miles from the coast. I spent two days there with a white trader who had been stationed there for a year, and I was assured by him that there were no gorillas known in that part. The natives report that they have been found in the lowlands south of there in the direction of the Ogowe basin; but their reports are conflicting, and none of them, so far as I could learn, claim that he is found north of there, nor in the mountains eastward. I admit the possibility that he has been found and may yet inhabit the strip of land between this river and the Ogowe, but I repeat that there is no proof that he was ever found north of the Gaboon. With due respect to Sir Richard Owen and others who have never been in that country, I insist that they are mistaken.

It is true that one of the tribes living north of the Gaboon has a name for this animal, but it does not follow that he lives in that country. The Orunga tribe have a name for lion, but there is not such a beast within 400 miles of their country, and not one of that tribe ever saw one.

A vast number of specimens have been secured at Gaboon, but they have been brought there from far away, because it is the chief town of the colony, and there are more white men there to buy them than elsewhere. It is quite impossible for a stranger to ascertain what part a specimen is brought from. The native hunter will not tell the truth lest some one else should find the game and thus deprive him of its capture and sale.

I once saw a specimen at Cameroon, and was told that it had been captured in that valley fifty miles from the coast; but I hunted up its history and found with absolute certainty that it was captured near Mayumba, 200 miles south of Gaboon. Even with the greatest care in hunting up the history of specimens one may fail, and often does in tracing it to its true source, but every one so far, that I have followed up, has been brought somewhere within the limits I have laid down. Contrary to the statement of some authorities that these apes "have never been seen on the coast" since 1852, the greatest number of them are found near the coast. I do not mean to say that they sit on the sand along the beach, or bathe in the surf, but they live in the jungle of that part.

Along the Lower Congo the gorilla is known only in name, and scores of the natives do not know even that. The nearest point to that river that I have been able to locate the gorilla as a native, is in the territory about sixty or seventy miles north-west of Stanley Pool.

I am indebted to the late Carl Steckelmann, who was drowned at Mayumba in my presence last October. He was an old resident of the coast, a good explorer, a careful observer, and an extensive traveller. I knew him well, and secured from him much information concerning the gorilla. He traced out with me, on a map, what he believed to be the south and south-east limits of the gorilla. Not thirty minutes before the fatal accident in which he lost his life, I had closed arrangements with him to make an expedition from Mayumba to the Congo, near Stanley Pool, by one route, and return by another, but his death prevented its fulfilment.

Dr. Wilson, who was the first missionary at Gaboon and located there in 1842, wrote a lexicon of the native language about six years after that time. In this he entirely omits the name of the gorilla. Dr. Walker eight years later gave the definition, "a monkey larger than a man." But he had never seen a specimen of the ape, except the skulls and a skeleton which were brought from other parts. It is true that Dr. Savage first learned at Gaboon about the gorilla, and secured a skull at that place from which he made drawings, and on which account his name was attached to the animal in Natural History. Dr. Ford a few years later sent the first skeleton to America, and Captain Harris sent the first to England. The former is in the Museum of Zoology at Philadelphia. Both of these specimens may have come from any place a hundred miles away from Gaboon.

It is possible at this early date the gorilla may have occupied the peninsula south of the Gaboon River, in greater numbers than he has ever done since, because up to that time there had been no demand for him; but if such was true at that time, it is not so now, and if he is not extinct in that part, he is so rare as to make it doubtful whether or not he is found there at all.

In four journeys along the Ogowe River and the lakes of that valley, I made careful inquiries at many of the towns, and the natives assured me that the gorillas lived on the south side of that river. I spent five days at the village of Mbiro, which is located on the north side of the river and about fifty miles from the coast. There I was told by the native woodsmen that no gorillas lived on the north side, but there were plenty of them along the lakes south of the river. They said that in the forest back of that town were plenty of chimpanzees, and that they were sometimes mistaken for gorillas, but there were absolutely none of the latter in that part. In view of these and countless other facts, I deem it safe to say that few or no gorillas can be found north of the Ogowe River at any point, and I even doubt if the specimen heard of on the Komo was a genuine gorilla. The natives sometimes claim to have something of the kind for sale in order to get a bonus from some trader, when in truth he may not have anything of the kind.

The only point north of the Ogowe at which I had any reason to believe a gorilla could be found was in the neighbourhood of a small lake called Inenga. This lake is nearly due west from the mouth of the Nguni River and something more than a hundred miles from the coast. Certain reports along that part appeared to have some flavour of truth, but there was no proof except the word of the natives.

In the lake region south of the river they are fairly abundant as far south as the head-waters of the Rembo Nkami and through the low country of the Esyira tribe, but they are very rare in the forests, and unknown in the highlands and plains of this country. South of the Chi Loango they are quite unknown, and south of the Congo never heard of.

There are no means possible to estimate their number, but they are not so numerous as may be supposed, and from the reckless slaughter of them by the natives in order to secure them for white men, they may soon become extinct. Their ferocity alone has saved them up to this time from such a fate, but the use of approved arms will soon overcome that.

The skeleton of the gorilla is so nearly the same as that of the chimpanzee, which has elsewhere been compared to the human skeleton, that we shall not review the comparison at length, but must note one marked feature in the external form of the skull, which differs alike from other apes and man.

The skull of the young gorilla is much like that of the chimpanzee, and remains so until he approaches the adult state; but as he approaches this period, the ridge above the eyes becomes more prominent, and at the same time a sharp, bony ridge begins to develop along the temples, and continues around the back of the head on that part of the skull called the occiput. At this point it is intersected by another ridge at right angles to it. This is called the sagittal ridge, and runs along the top of the head towards the face; but on the forehead it flattens nearly to the level of the skull, and divides into two very low ridges, which turn off to a point above the eyes and merges into that ridge. These appear to be a continuous part of the skull, and are not joined to it by sutures. The mesial crest in very old specimens rises to the height of nearly two inches above the surface of the skull, and imparts to it a fierce and savage aspect; but in the living animal the crests are not seen, as the depressions between them are filled with large muscles, which make the head look very much larger than it would otherwise. These crests affect only the exterior of the skull, and do not appear to alter the form or size of the brain cavity, which is larger in proportion than that of the chimpanzee. These crests are peculiar to the male gorilla, and the female skull shows no trace of them.

PLATE I

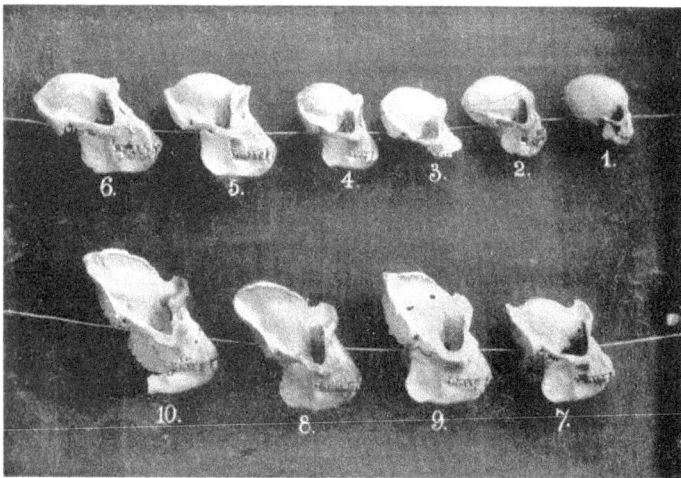

PLATE II

There is at least one case in which this crest has failed to develop in the male. By reference to the series of skulls found in the cuts given herewith,

No. 6 is that of an adult male, which I know to be such, as I dissected him and prepared the skeleton myself. He was killed in the basin of Lake Fernan Vaz, not more than two or three hours from my cage, and his body was brought to me at once. A good idea of his size can be obtained by reference to another cut given herewith, where I have some natives skinning him. In this picture he is sitting flat on the sand; his body is limp, and is somewhat shorter than it was in life, and yet it can be seen that the top of his head is higher than the hip of the man who is holding him. On the left of the gorilla, in the foreground, sits the man who killed him. He is sitting on a log, and it did not occur to me until too late to place them side by side in order to make a comparison. The body and head of this gorilla as he sits measured nearly four feet from the base of the spinal column to the top of the head. I did not weigh him, but made an estimate by lifting him in my hand, and believed he weighed at least 240 lbs. Yet he was not an old specimen, but if compared to No. 7, in which the crests are well developed, it is found to be larger, and other things point to the fact that he was older.

I am aware that one specimen of itself does not prove anything, but it shows in this case that this ape does not always develop that crest. His head was surmounted by the red crown which we have described, and No. 1, which is the skull of Othello, had the same mark. He was captured near the place where No. 6 was killed. No. 2, which is the skull of a young female nearly four years old, had the same, and she was also captured in the same basin, but on the opposite side of the lake.

The facial bones of No. 6 showed that the animal had received a severe blow in early life, but the fragments had knitted together, and the effect could not be seen in the face of the ape while alive. In this same picture it will be noticed that the lower lip hangs down so low that the mouth is opened. The lip is very massive and mobile, and in this character he resembles the negro. The lower lip is much thicker and more flexible than the upper.

No. 8 is the skull of a large male from Lake Izanga, which is on the south side of the Ogowe River, more than a hundred miles from the coast, and is one of the three centres of population mentioned. I do not know its history. It was presented to me by Mr. James Deemin, an English trader with whom I travelled many days in the Ogowe River; and I wish here to take occasion to express my sincere thanks to him for the many kindnesses extended to me.

No. 5 is the skull of an adult female. By comparing it in profile to No. 6 it will be seen that they resemble, but the muzzle of the latter projects a little more, and the curvature of the skull across the top is less: the distance a little greater.

Nos. 2, 3, 4 and 5 are female; the others are all male.

Nos. 3, 4, 5, 7, 9 and 10 belong to the Liverpool Museum, but are shown here for comparison. The other four are all at Toronto University.

While this series is not complete in either sex, it is an excellent one for comparative study.

I do not know whether the heads of those with the crests were the same colour as No. 6 or not, but the *ntyii*, which I have mentioned as possibly a new species of the gorilla, does not have this crown of red. His ears are also said to be larger than those of the gorilla, but smaller than the chimpanzee's, and he is reputed to grow to a larger size than either of them.

The skin of the gorilla is a dull black or mummy colour over the body, but that of the face is a jet black, quite smooth and soft. It looks almost like velvet.

One fact peculiar to this ape is, that the palms of both hands and feet are perfectly black. In other animals these are usually lighter in colour than the exposed parts. In all races of men, in all other apes, monkeys, baboons, and lemurs, the palms are lighter than the backs of the hands, and the same is true of the feet. The thumb of the gorilla is more perfect than that of the chimpanzee, yet it is smaller in proportion to the hand than in man. The hand is very large, but has more the shape of the hand of a woman than that of man. The fingers taper in a graceful manner, but appear much shorter, by reason of the web alluded to, than they really are. It is not really a web, in the true sense, but the integument between the fingers is extended down almost to the second joint, but the forward edge of the web, when the fingers are spread, is concave; when brought together, the skin on the knuckles becomes wrinkled, and the web almost disappears. This effect is more readily noticed in the living animal than in the dead. The texture of the skin in the palms is coarsely granulated, and the palmar lines are indistinct. The great toe sets at an angle from the side of the foot, like a thumb, but has more prehensile power than that of the hand; but the foot is much less flexible, and has less prehensile power.

At this point I desire to draw attention to one important fact. The tendons of the foot, which open and close the digits, are imbedded in the palm in a deep layer of coarse, gristly matter, which forms a pad, as it were, under the sole of the foot, and prevents it from bending; therefore it is not possible for the gorilla to sleep on a perch. In this respect he resembles man more than the chimpanzee does, but it is quite certain that neither of them have the arboreal habit. The gorilla is an expert climber, but cannot sleep in a tree. In the hand the tendons which close the fingers are the same length as the line of the bones, and this permits him to open the fingers to a straight line, which the chimpanzee cannot do.

One other important point I desire to mention. The muscles in the leg of a gorilla will not permit it to stand or walk erect. The large muscle at the back of the leg is shorter than the line of the bones of the leg above and below the knee; and when this muscle is brought to a tension, those bones form an angle of about 130 degrees, or thereabouts; and so long as the sum of two sides of a triangle is greater than the other side, a gorilla can never bring his leg into a straight line. In the infant state the muscle is pliant or elastic, and the bones less rigid, so that in that state it can be made nearly straight. The habit of hanging by the arms and walking with them in a straight line develops the corresponding muscle in that member, so that the bones can be brought in line.

The gorilla can stand upon his feet alone, and walk a few steps in that position; but his motion is awkward, because his knees turn outward, forming an angle of 30 or 35 degrees on either side of the mesial plain. He never attempts to walk in this position, except at perfect leisure, and then usually holds on to something with his hands. The tallest gorilla known, when perfectly erect, is about 6 feet 2 inches.

The leg of the gorilla from the knee to the ankle is almost the same size. In the human leg there is what is called the "calf" of the leg, but this in the apes is very small; however, there is a slight tendency in that direction, and it must be noted that in the human species the calf of the leg appears to belong to the higher types of men; and as we descend from the highest races of mankind this character disappears as we approach the savage. The pigmies and the bushmen have the smallest of any other men. It is not to be inferred from this that apes would ever have this feature developed in them by elevating them to a higher plane so long as they remained apes; but it is possible that such a result would follow in the course of time.

One thing which tends to lessen this in the gorilla is the size of the muscles about the ankle and the flexibility of that joint. Also the joint of the knee, being much larger in proportion to the leg, makes the calf appear smaller than it really is.

The corresponding part of the arm is more like that part of the human body.

In a sitting posture the gorilla rests his body upon the ischial bones, with his legs extended or crossed, while the chimpanzee usually squats, resting those bones upon his heels. He sometimes sits, but more frequently squats. When in these attitudes, both usually fold their arms across their breasts.

The hair of the gorilla is irregular in growth. It is more dense than that of the chimpanzee, but less uniform in size and distribution. On the breast it is very sparse, on the arms, long, and on the back, dense, and interspersed with long coarse hairs. The ground of colour is black, but the extreme end of the hair is tipped with pale white. This is so in early youth, and with age the white encroaches, until, in extreme age, the animal is quite grey. The top of the head is covered with a thick growth of short hair, of a dark tan colour, which looks almost like a wig. This mark seems to be peculiar to certain localities, but is uniform among those captured in the Fernan Vaz basin.

YOUNG GORILLA WALKING

A white trader living on this lake claims to have seen a gorilla which was perfectly white. It was seen on the plain near the lake. It was in company with three or four others. It was thought to be an albino, but in my opinion it was only a very aged specimen turned grey. A few of them have been secured that were almost white. It is not, however, such a shade of white as would be found in an animal whose normal colour is white. I cannot vouch for the colour of this ape seen on the plain, but there must have been something peculiar in it to attract so much attention among the natives.

So far, only one species of this ape is known to science, but there are reasons to believe that two species exist. In the forest regions of Esyira the natives described to me another kind of ape, which they averred was a half-brother to the gorilla. They know the gorilla by the native name *njina*, and the other type by the name *ntyii*. They did not confuse this with the native name *ntyigo*, which is the name of the chimpanzee, nor with *kulu-kamba*, all of which are known to them; but they described in detail, and quite correctly, the three known kinds of ape, and in addition gave me a minute account of the appearance and habits of the fourth kind, which I believe to be another species of the gorilla. They claim that he is more intelligent and human-like than any one of the others; and they say that his superior wisdom makes him more alert, and therefore more difficult to find. He is said always to live in parts of the forest most remote from human habitation.

The dental formula of the gorilla is the same as that of man, but the teeth are larger and stronger, and the canine teeth are developed almost into huge tusks. One thing to be remarked is the great variety of malformations in the teeth of this animal. It is a rare thing to find among them a perfect set of teeth, except in infancy. The cause of this appears to be violence or accident.

The eyes of the gorilla are large, dark, and expressive, but there is no trace of white in them. That part of the eye which is white in man is a dark coffee-brown in the gorilla, but becomes lighter as it approaches the base of the optic nerve. The taxidermist or the artist, who often furnishes him with a white spot in the corner of his eye, does violence to the subject; and those who pose the animal with his mouth open like a fly-trap, and his arms raised like a lancer, ought to be banished from good society. It is true that such things lend an aspect of ferocity to the creature, but they are caricatures of the thing they mean to portray.

The ears of the gorilla are very small, and lie close to the sides of the head. The model of them is much like the human ear.

I shall not pursue the comparison into minute details, but leave that to the specialist, in whose hands it will be treated with more skill and greater scope. As my especial line of research has been in the study of their speech and habits, I shall confine myself to that, but the general comparison I have made is necessary to a better understanding of the subject.

CHAPTER XV
HABITS OF THE GORILLA

A study of the habits of the gorilla in a wild state is attended with much difficulty, but the results that I obtained during a sojourn of one year among them are an ample reward for the efforts made. In a state of captivity the habits of animals are made to conform in a measure to their surroundings, and since those are different many of their habits differ also. Some are foregone, others modified, and new ones acquired, therefore we cannot know with certainty what the animal was in a state of nature. In the social life of the gorilla there are a few things perhaps that differ very much from that of the chimpanzee, but there are some that do in a certain degree. From the native accounts of the modes of life of these two apes, there would appear to be a much greater difference than a systematic study of them reveals; but the native version of things frequently has a germ of truth which may serve as a clue to the facts in the case; and while we cannot rely upon the tales they relate in all details, we can forgive the mendacity and make use of the suggestion they furnish.

It is certain that the gorilla is polygamous in habit, and it is probable that he has an incipient idea of government. Within certain limits he has a faint perception of order and justice, if not of right and wrong. I do not mean to ascribe to him the highest attributes of man, or exalt him above the plane to which his faculties assign him; but there are reasons to justify the belief that he occupies a higher social and mental sphere than other animals, except the chimpanzee.

In the beginning of his career, in independent life, the gorilla selects a wife with whom he appears to sustain the conjugal relations thereafter, and preserves a certain degree of marital fidelity. From time to time he adopts a new wife, but does not discard the old one; in this manner he gathers around him a numerous family, consisting of his wives and their children. Each mother nurses and cares for her own young, but all of them grow up together as the children of one family. There is no doubt that the mother sometimes corrects and sometimes chastises her young, which suggests a vague idea of propriety. The father exercises the function of patriarch in the sense of a ruler, and the natives call him *ikomba njina*, which means gorilla king. To him the others all show a certain amount of deference. Whether this is due to fear or to respect, however, is not certain, but here is at least the first principle of dignity.

The gorilla family, consisting of this one adult male and a number of females and their young, are within themselves a nation. There do not appear to be any social relations between different families, but within the same household there is apparent harmony.

The gorilla is nomadic, and rarely ever spends two nights in the same place. Each family roams about in the bush from place to place in search of food, and wherever they may be when night comes on they select a place to sleep and retire. The largest family of gorillas that I have ever heard of was estimated to contain twenty members. But the usual number is not more than ten or twelve. The chimpanzee appears to go in larger groups than these, and sometimes in a single group two or even three adult males have been seen. When the young gorilla approaches the adult state, he leaves the family group, finds himself a mate, and sets out in the world for himself. I observed that, as a rule, when one gorilla was seen alone in the forest it was usually a young male, but nearly grown; it is probable that he was then in search of a wife. At other times two only are seen together, and in this event they are usually a pair of male and female, and generally young. Again, it sometimes occurs that three adults are seen with two or three children; often one of the children two or three years old, and the others a year younger, which would indicate that the male had had one of his wives much longer than the other. In large families young ones of all ages, from one year old to five or six years old, are seen; but the fact is plain that the older children are much fewer in number. I have once seen a large female with her babe, quite alone; whether she lived alone or was only absent for the moment I cannot tell.

The king gorilla does not provide food for his family, but, on the contrary, it is said they provide for him. I have been informed on two occasions, from different sources, that the king gorilla has been seen sitting quietly under the shade of a tree, eating, while the others collected and brought to him the food. I have never witnessed such a scene myself, but it does not seem probable that the same story would have come from two sources unless there was some foundation for it.

In the matter of government, the gorilla appears to be somewhat more advanced than most animals. He leads the others on the march, and selects their feeding grounds and places to sleep; he breaks camp, and the others all obey him in these respects. Other animals that travel in groups do the same thing; but in addition to this, the natives aver that the gorillas from time to time hold palavers or a rude form of court or council in the jungle. On these occasions, it is said the king presides; that he sits alone in the centre, while the others stand or sit in a rough semicircle about him, and talk in an excited manner. Sometimes the whole of them are talking at once, but what it means or alludes to no native undertakes to say, except that it has the nature of a quarrel. To what extent the king gorilla exercises the judicial function is a matter of grave doubt, but there appears to be some real ground for the story.

As to the succession of the kingship there is no certainty, but the facts point to the belief that on the death of the king, if there be an adult male he assumes the royal prerogative, otherwise the family disbands, and they are absorbed by or attached to other families. Whether this new leader is elected in the manner that other animals appoint a leader, or assumes it by reason of his age, cannot be said; but there is no doubt that in many instances families remain intact for a time after the death of their leader.

It has been said by many that the gorilla builds a rude hut or shelter for himself and family, but I have found no evidence that such is true. The natives declare that he does so, and some white men affirm the same; but during my travels through their habitat, I offered liberal and frequent rewards to any native who would show me one of these specimens of simian architecture, but I was never able to find any trace of one made or occupied by any ape. They may sometimes, and doubtless do, take shelter from the tornadoes, but it is always under some fallen tree or cluster of broad leaves, and there is nothing to show that they arrange any part of them. So far as I could find, there is no proof that any gorilla ever put two sticks together with the idea of shelter. As to his throwing sticks or stones at an enemy, I have found nothing to verify it; in my opinion, it is a mere freak of fancy.

The current opinion or idea that a gorilla will attack a man without being provoked to it, is an error. He is shy and timid, and shrinks alike from man and other large animals. I have no doubt that when he is in a rage he is both fierce and powerful, but his ferocity and strength are rated above their true value. In combat he is a stubborn foe no doubt, but no one that I have met has ever seen him thus engaged.

The mode of attack as described by many travellers is a mere theory. It is said in this act he walks erect, beats with fury on his breast, roars and yells, and in this manner seizes his adversary, tears open his breast, and drinks the blood. I have never seen a large gorilla in the act of assault. During the time of my stay in the jungle I had a young gorilla in captivity, and I made use of him in studying the habits of his race. I kept him tied with a long line which allowed him room to play and climb, and at the same time prevented him from escaping into the forest, which he always tried to do the instant he was released. I released him frequently for the purpose of watching his mode of attack when recaptured. While being pursued he rarely looked back, but when overtaken he invariably assailed his captor. This gave me an opportunity of seeing his method of attack, in which he displayed both skill and judgment. As my boy would approach him, he would calmly turn with one side to the foe and, without facing the boy, would roll his eyes in such a manner as to see him and at the same time conceal his purpose. When the boy came within reach, the gorilla would grasp him with a thrust of the arm to one side and slightly backward. When he had seized his adversary by the leg, he would instantly swing the other arm round with a long sweep and strike the boy a hard blow; then he began to use his teeth. He seemed to depend more upon the blow than the grasp, but the latter served to hold the object of attack within reach; in every case he kept one arm and one leg in reserve until he had seized his adversary. It is true that these attacks were made upon an enemy in pursuit, but his mode appeared to be a normal one; he could strike a severe blow, and did not show any sign of tearing or scratching his opponent. In these attacks he made no sound of any kind. I do not pretend to say that other gorillas do not scream or tear their victims, but I take it that the habits of the young are much, if not quite, the same as those of their parents, and from a study of this specimen I am forced to modify many opinions imbibed from reading or from pictures and specimens which I have seen. Many of them represent the gorilla in absurd and sometimes impossible attitudes. They certainly do not represent him as I have seen him in his native wilds.

When the chimpanzee attacks, so far as I have seen among my own specimens, he approaches his enemy and strikes with both hands, one slightly in advance of the other. After striking a few blows, he will grasp his opponent and use his teeth, then shoving him away again uses his hands, and usually, on beginning the attack, accompanies the assault with a loud, piercing scream. Neither he nor the gorilla closes the hand to strike, nor uses any weapon except the hands and teeth. I had another young female gorilla for a short time as a subject for study. Her mode of attack appeared to be the same, but she was too large to risk in such experiments.

I have read and heard descriptions of the sounds made by the gorilla, but nothing ever conveyed to my mind an adequate idea of their true nature, until I heard them myself within a few hundred feet of my cage in the dead of night. By some it has been called roaring, and by others howling; but it is neither truly a roar nor a howl. They utter a peculiar combination of sounds, beginning in a low, smooth tone, which rapidly increases in pitch and frequency, until it becomes a terrific scream. The first part of the series is quite within the scope of the human voice, but as it rises in pitch and increases in volume it passes far beyond the reach of the human lungs. The first sound of the series and each alternate sound is made by expiration, while the intermediate ones appear to be by inspiration, but how it is accomplished is difficult to say. The sound as a whole resembles the braying of an ass, except the notes are shorter, the climax higher, and the sound is louder. A gorilla does not yell in this manner every night, but when he does so it is usually between two and five o'clock in the morning; I have never heard the sound during the day nor in the early part of the night. When he thus screams, he repeats the series from ten to twenty times, at intervals of one or two minutes each. I know of nothing in the way of vocal sounds that can inspire such terror as the voice of the gorilla. It can be heard over a distance of three or four miles. I could assign no definite meaning to it unless it was intended to alarm some intruder that came too near.

One morning between three and four o'clock I heard two of them screaming at the same time. I do not mean to say at the same instant, but at intervals during the same period of time. One of them was within about a third of a mile of me, and the other in another direction perhaps a mile away. The points we occupied respectively formed a scalene triangle. The sounds did not appear to have any reference to each other. Sometimes they would alternate, and at other times they would interrupt each other. They were both made by giants of their kind, and every leaf in the forest vibrated with the sound. This was during the latter part of May. They do scream in this way from time to time throughout the year, but it is most frequent and violent during February and March.

This wild screaming is sometimes accompanied by a peculiar beating sound. It has been described by travellers, and currently believed to be made by the animal beating with his hands upon his breast; but such is not the case. It is very certain that the sound cannot be made by that means. The quality of the sound shows that such cannot be the means employed. I have heard this beating several times, and have paid marked attention to its character. At a great distance it would be difficult to discern the exact quality; but on one occasion, while stopping over-night in a native town, I was aroused from sleep by a gorilla screaming and beating within a few hundred yards. I put on my boots, took my rifle, and cautiously crossed the open ground between the village and the forest. This brought me within about two hundred yards of the animal. The moon was faintly shining, but I could not see the beast, and I had no desire to approach nearer at such a time, but I heard distinctly every stroke. I believe the sound was made by beating upon a log or piece of dead wood. He was beating with both hands, the strokes alternating with great rapidity, and not unlike the manner in which the natives beat a drum, except that the hand made the same number of strokes, and the strokes were in a constant series, rising and falling from very soft to very loud, and *vice versâ*. A number of these runs followed one another during the time the voice continued. Between the first and second strokes the interval was slightly longer than that between the second and third, and so on through the scale. As the beating increased in loudness the interval shortened in an inverse degree, while in descending the scale the intervals lengthened as the beating softened, and the author of the sound was conscious of this fact. I could trace no relation in time or harmony between the sound of the voice and the beating, except that they began at the same time and ended at the same time. The same series of vocal sounds was repeated each time, beginning on the low note and ending on the highest note or pitch in each case, while the rise and fall of the series of the beaten sounds was not measured by the duration of the voice. The series each time began with a soft note, but ended at any part of the scale at which the voice ceased, and was not the same in every case.

NATIVE CARRIER BOY

I have no doubt that the gorilla beats upon his breast: he has been seen to do so in captivity, but the sounds described above were not so made. Since the gorilla makes these sounds only at night, it is not probable that any man ever saw him in the act. It does not require a delicate sense of hearing to distinguish a sound made by beating the breast from that of dead wood or other similar substance.

I have attributed the above sound to the gorilla, because I have been assured by many white men and scores of natives that it was made by him; but since my return from Africa I have had time to consider and digest certain facts tabulated on that trip, and as a result I am led to doubt whether this sound is made by the gorilla or not. There are good reasons to believe that it is made by the chimpanzee instead, and I shall state them.

I observed that my own chimpanzees made this sound exactly the same as that I heard in the forest, except that it was less in volume, which was due to their age. I could induce them at any time to make the sound, and frequently did so in order to study it. On my arrival in New York I found that Chico, the big chimpanzee belonging to Mr. Bailey, frequently made the same sound at night. It was said to be so loud and piercing that it fairly shook the stately walls of Madison Square Garden. From reading the description given by the late Professor Romanes of the sound made by "Sally" in the London Gardens, it appears to be the same sound.

It is well known to the natives that the chimpanzees beat on some sonorous body, which they call a drum. Four years ago I called attention to the habit of the two chimpanzees in the Cincinnati Gardens. They frequently indulged in beating upon the floor of their cage with their knuckles. This was done chiefly by the male. The late E. J. Glave described to me the same thing, as being done by the chimpanzees in the Middle Congo basin.

It is not probable that two animals of different genera utter the same exact sound, and this is more especially true of a sound that is complex or prolonged. Neither is it likely that the two would have a common habit, such as beating on any sonorous body. Since it is certain that one of these apes does make the sound described, it is more than probable that the other does not. The same logic applies to the beating.

Many things that are known of the chimpanzee are taken for granted in the gorilla, but it is erroneous to suppose that in such habits as these they would be identical. In some cases I have been able to prove quite conclusively that the chimpanzee alone did certain things which were ascribed to the gorilla.

In view of these facts alone, I am inclined to believe that after all, the sound described is made by the chimpanzee and not by the gorilla.

Another case in which the gorilla is portrayed is wrong. The female gorilla is represented as carrying her young clinging to her waist. I have seen the mother in the forest with her young mounted upon her back, with its arms around her neck and its feet hooked in her armpits. I have never seen the male carry the young, but in a number of specimens of advanced age I have seen a mark upon the back and sides which indicates that he does so. It is in the same place that the young rest upon the back of the mother. In form it is like an inverted **Y**, with the base resting on the neck and the prongs reaching under the arms. This mark is not one of nature, but appears to be the imprint of something carried there. In a few specimens the hair is worn off until the skin is almost bare. The prongs are more worn than the stem of the figure, which is due to the fact that more weight is borne upon those parts than elsewhere. I do not assert that such is the cause, but it is worthy of note that such is the fact.

The gorilla is averse to human society. He is morose and sullen in captivity. He frets and pines for his liberty. His face appears to be incapable of expressing anything like a smile, but when in repose it is not repugnant. In anger his visage depicts the savage instincts of his nature. The one which lived with me for a time in the forest was a sober, solemn, stoical creature, and nothing could arouse in him a spirit of mirth. The only pastime he indulged in was turning somersaults. Almost every day, at intervals of an hour or so, he would stand up for a moment, then put his head upon the ground, turn over like a boy, rise to his feet again, and look at me as if expecting my applause. He would frequently repeat this act a dozen times or more, but never smiled or evinced any sign of pleasure. He was selfish, cruel, vindictive, and retiring.

One peculiar habit of the gorilla, both wild and in captivity, is that of relaxing the lower lip when in repose. They drop the lid until a small red line appears across the mouth from side to side. It is not done when in a sullen mood, but when perplexed or in a deep study.

Another constant habit is to protrude the end of the tongue between the lips, until it is about even with the outer edge of them. The end of the tongue is somewhat more blunted than that of the human. This habit is so frequent with the young gorilla that it would appear to have some meaning, but I cannot suggest what it is.

The habit of the gorilla, in sleeping, is to lie upon the back or side, with one or both arms placed under the head as a pillow. He cannot sleep on a perch, as we have already noted, but lies upon the ground at night. I had once pointed out to me the place at the base of a large tree where a school of them had slept the night before. One imprint was quite distinct. The stories told about the king gorilla placing his family in a tree while he sits on watch at the base, is another case of supposition.

A YOUNG GORILLA ASLEEP

The food of the gorilla is not confined to plants and fruits. They are fond of meat, and eat it either raw or cooked. They secure a small supply by catching rodents of various kinds, lizards and toads; they are also known to rob the nests of birds of the eggs, and of the young. A native once pointed out to me the quills and bones of a porcupine which he said had been left by a gorilla who had eaten the carcass, and he said that it was not at all rare for them to do so. The fruits and plants they live upon chiefly are acidulous in taste, and some of them are bitter. They often eat the fruit of the plantain, but prefer the stalk, which they twist and break open and eat the succulent heart of the plant. They do the same with the *batuna*, which grows all through the forest. The fruit of this plant is a red pod filled with seeds imbedded in a soft pulp, it is slightly acidulate and astringent. The wild mangrove which forms a staple article of food for the chimpanzee is rarely, if ever, touched by the gorilla, and the same is true of many other plants and fruits. I once saw a gorilla try to seize a dog, but whether it was for the purpose of eating the flesh or not I cannot say. One, however, did catch and devour a small dog on board the steamer *ship*, while on a voyage home from Africa. Both belonged to Captain Button, who assured me of the fact. They have no fixed hours for eating, but usually do so in the early morning or late afternoon. I have, in a few instances, seen them refuse meat. They are perhaps less devoted to eating flesh than the chimpanzee.

In the act of drinking, the gorilla will take a cup, place the rim in his mouth and drink like a human being. He does this without being taught, while the chimpanzee prefers to put both lips in the vessel. I have never known one that would drink beer, spirits, coffee or soup, but their drink is limited to milk or water, while the chimpanzee drinks beer and other things as well.

NATIVE WOMEN OF THE INTERIOR

CHAPTER XVI
OTHELLO AND OTHER GORILLAS

While I was living in my cage in the jungle I secured a young gorilla, to whom I gave the name "Othello." He was about one year old, strong, hardy and robust. I found him to be a fine subject for study, and made the best use of him for that purpose. I have elsewhere described his character, but his illness and death are matters of profound interest.

At noon on the day of his decease he was quite well and in fine humour. He was turning somersaults and playing like a child with my native boy. In his play he evinced a certain interest, and his actions indicated that it gave him pleasure, but his face never once betrayed the fact. It was amusing to see him with the actions of a romping child and the face of a cynic.

He was supplied with plenty of native food, had a good appetite, and ate with a relish. Just after noon I sent the boy on an errand, and he was expected to return about night. Near the middle of the afternoon I observed that Othello was ill; he declined to eat or drink, and lay on his back on the ground, with his arms under his head as a pillow. I tried to induce him to walk with me, to play, or to sit up, but he refused to do so. By four o'clock he was very ill. He rolled from side to side, and groaned as if in pain. He kept one hand upon his stomach, where the pain appeared to be located. He displayed all the symptoms of gastric poisoning, and I have reason to believe now that the boy had given him poison. I should regret to foster this suspicion against an innocent person, but it is based upon certain facts that I have learned since that time.

While I sat in my cage watching Othello, who lay on the ground a short distance away, I discovered a native approaching him from the jungle. The man had an uplifted spear in his hand, as if in the act of hurling it at something. He had not seen me, but it did not for the moment occur to me that he had designs upon my pet. I spoke to him in the native language, when he explained that he had seen the young gorilla, and from that fact suspected there was an old one close at hand, for whose attack he was prepared: that he was not afraid of the little one, but desired to capture it. I informed him that my gorilla was ill. He examined it, and assured me that it would die. The man departed, and Othello continued to grow worse. His sighing and groaning were really touching. I gave him an emetic, which took effect with good results. I also used some vaperoles to resuscitate him, but my skill was not sufficient to meet the demands of his case.

His conduct was so like that of a human being that it deeply impressed me, and being alone with him in the silence of the dreary forest at the time of his demise, gave the scene a touch of sadness that impressed me with a deeper sense of its reality; and Moses watched the dying ape as if he knew what it meant. He showed no signs of regret, but his manner was such as to suggest that he knew it was a trying hour.

Othello died just before sunset, but for a long time prior to this he was unconscious. The only movements made by him were spasmodic actions of the muscles caused by pain. The fixed and vacant stare of his eyes in this last hour was so like those of man in the hour of dissolution, that no one could look upon the scene and fail to realise the solemn fact that this was death. The next day I dissected him, and prepared the skin and skeleton to bring home with me. They are now, with Moses and others, in the Museum of the University of Toronto; and if the taxidermist who mounts the skin of Othello poses him like most of the craft do—in the attitude of dancing a fandango and the corners of his mouth forming obtuse angles—I will have that man executed if I have to bribe the court.

When I first secured this ape and brought him to my home in the bush, he was placed on the ground a few feet from my cage, and near him was laid some bananas and sugar-cane for Moses, who had not yet seen the stranger. The gorilla was in a box with one side open, so that he could easily be seen. My purpose was to see how each one would act on discovering the other. When Moses observed the food he proceeded to help himself. On seeing the gorilla he paused a moment and gave me an alarm, but he was not himself deterred from taking a banana, which he seized and retreated. While he was eating the banana, I took the gorilla from the cage and set him on the ground by it. I petted him, and gave him some food. Moses looked on, but did not interfere. I returned to my cage, and Moses proceeded to investigate the new ape. He approached slowly and cautiously within about three feet of it. He walked around it a couple of times, keeping his face towards it, and gradually getting a little nearer. At length he stopped by one side of the gorilla, and came up within a few inches of it. He appeared to stand almost on tiptoe, with only the ends of his fingers touching the ground. The gorilla continued to eat his food without so much as giving him a look. Moses placed his mouth near the ear of the gorilla and gave one terrific yell. But the gorilla did not flinch or even turn his eyes. Moses stood for a moment looking at him as if in surprise that he had made no impression. After this time he made many overtures to make friends with the gorilla, but the latter did not entertain them with favour beyond maintaining terms of peace. They never quarrelled, but Othello always treated him as an inferior. I do not know if he entertained a real feeling of contempt, but his manner was such.

There were but few articles of food that he and Moses liked in common, and therefore they had no occasion to quarrel; but they never played together or cultivated any friendly terms as the chimpanzees did among themselves. This may have been due to the gorilla, who was so exclusive in his demeanour towards the chimpanzee as to forbid all attempts of the latter to become intimate. The chimpanzee by nature is more sociable and is fond of human society. He imitates the actions of man in many things, and quickly adapts himself to new conditions, while the gorilla is selfish and retiring. He can seldom, if ever, be reconciled to human society; he does not imitate man nor yield to the influences of civilised life.

One special trait of the gorilla which I wish to emphasise is that he is one of the most taciturn, if not quite the most, of any member of the simian family. This fact does not appear to confirm my theory as to their high type of speech, but it is a fact so far as I observed, although the natives say that they are as loquacious as the chimpanzee. Among the specimens that I have studied, both wild and in captivity, I have never heard but four sounds that differed from each other, and of these only two could properly be defined as speech. I do not include the screaming sound described in another chapter. I have not been able so far to translate the sounds that I have heard, and they cannot be spelled with letters. There is one sound which Othello often used. It was not a speech sound, but a kind of whine, always coupled with a deep sigh. When left alone for a time he became oppressed with solitude. At such times he would heave a deep sigh and utter this strange sound. The tone and manner strongly appealed to the feelings of others, and while he did not appear to address it to any one or have any design in making it, it always touched a sympathetic chord, and I was sometimes tempted to release him. Another sound which was not within the pale of speech was a kind of grumbling sound. This frequently occurred when he was eating. It was not a growl in the proper sense, but was in a way a kind of complaint. Twice I heard this same sound made by wild ones in the forest near my cage. The only thing that I can compare it to in its use is that habit of a cat while eating, to make a peculiar growling sound, which appears to be done only when something else is near. It is possibly intended to deter others from trying to take the food.

During my life in the cage I saw a number of gorillas, but I shall only describe a few of them, as their actions were similar in most instances.

The first one that I had the pleasure of seeing in the jungle came within a few yards of the cage before it was yet in order to receive. He was not half grown. He must have been attracted by the noise made in putting it together. He advanced with caution, and when I discovered him he was peering through the bushes as if to ascertain the cause of the sounds. When he saw me, he only tarried a few seconds and hurried off into the jungle. I did not disturb or shoot at him, because I desired him to return.

On the third day after I went to live in the cage a family of ten gorillas was seen to cross an open space along the back of a patch of plantains near one of the villages. A small native boy was within about twenty yards of them when they crossed the path in front of him. A few minutes later I was notified of it, took my rifle, and followed them into the jungle until I lost the trail. A few hours after this they were again seen by some natives not far away from my cage, but they did not come near enough to be seen or heard. The next day there was a family came within some thirty yards of the cage. The bush was so dense that I could not see them, but I could distinguish four or five voices. They seemed to be engaged in a broil of some kind. I suppose it was the same family that had been seen the day before. The second night after this time I heard the screams of one in the forest some distance from me, but I do not know whether it was the king of this family or another.

One day, as I sat alone, a young gorilla, perhaps five years old, came within six or seven yards of the cage and took a peep. I do not know whether he was aware of its being there or not until he was so near. He stood for a time, almost erect, with one hand holding on to a bough; his lower lip was relaxed, showing the red line mentioned above, and the end of his tongue could be seen between his parted lips. He did not evince either fear or anger, but rather appeared to be amazed. I heard him creeping through the bush a few seconds before I saw him, but as a rule they move so stealthily as not to be heard. I know of no other animal of equal weight that makes so little noise in going through the forest. During the short time he stood gazing at me I sat still as a statue, and I think he was in doubt as to whether I was alive or not. He did not turn and run away, but after a brief pause turned off at an angle and departed. He lost no time, but made no great haste. The only sound he made was a low grunt, and this he did not repeat.

At another time I heard two making a noise among the plantains near me. I could only obtain a glimpse of them, but as well as I could see they were of good size, being almost grown. They were making a low sound from time to time, something like I have described, but I could not see them well enough to frame any opinion as to what it meant. They were certainly not quarrelling, and I am not sure that they were eating, for I afterwards went and looked to see if I could find where they had broken any of the stalks. Their trail was visible through the grass and weeds, but I could find no stalk broken. They were moving at a very leisurely gait, and must have been within hearing ten or twelve minutes. They were quite alike in colour, and appeared to be so in size, although it is well known that the adult male attains a much greater size than the female.

On one occasion when I was standing outside of the cage some twenty yards away, Moses was sitting on a dead log near by. I turned to him, and was just in the act of sitting down by him when he gave an alarm. I looked around, and discovered a gorilla standing not more than twenty yards away. He had just that moment discovered us. He gazed for a few moments and started on, moving obliquely towards the cage. I turned to retreat. At this instant Moses gave one of his piercing screams, which frightened the gorilla and he fled. He changed his course almost at right angles. He was going at a good rate before Moses screamed, but he mended it at once.

One day I heard three sounds which my boy assured me were gorillas; they were in different directions from the cage. It was not a scream nor a howl, but somewhat resembled the human voice calling out with a sound like "he-oo!" This sound was repeated at intervals, but did not appear to be in the relation of call and answer, and the animals making them did not approach each other while doing so. The sounds were the same except in volume, and one of them appeared to be made by a much larger animal than the other two. I must say that this sound rarely occurred within my hearing during all my stay in that part, and with the exception of this time I never heard them make any loud sound during the day.

Another interesting specimen that I saw came prowling through the jungle as if he had lost his way. He found a small opening, or tunnel, which I had cut through the foliage in order to get a better view. Turning into that, he came a few steps towards the cage before he discovered it. Suddenly he stopped, squatted on the ground, but did not sit flat down. For a few seconds he was motionless, and so was I. He slowly raised one arm till his hand was above his head, in which position he sat for a few seconds, when he moved his hand quickly forward as if to motion at me. He did not drop his hand to the ground, but held it at an angle from his face for a short time, then slowly let it down till it reached the ground. During this time he kept his eyes fixed on me. At length he raised the other arm and seized hold of a strong bush, by which he slowly drew himself in a half-standing position. Thus he stood for a few seconds, with one hand resting on the ground. Suddenly he turned to one side, parted the bushes, and instantly disappeared. He uttered no sound whatever.

Another visitor that came within about thirty yards along the open path which led to my retreat, stopped when he discovered me, and stared in a perplexed manner. He turned away to retreat, but only went a few feet, turned around, and sat down on the ground. He remained in that attitude for more than half a minute, when he arose and retired in the direction from which he came.

The finest view that I ever had of any specimen, and at the same time the best subject for study, was a large female that came within a trifle more than three yards of me. There was a dog that belonged to a village a mile or two away that had become attached to me, and had found its way through the bush to my cage. He frequently came to visit me in my retreat, and I was always glad to welcome him. One afternoon, about three o'clock, he came, and I let him in the cage for a while to pass the usual greetings. I had a bone of a goat which I had saved from my last meal, and I threw this out to him in the bush a few feet away from the cage. He seized the bone, and began to gnaw it where it lay. His body was in the opening of a rough path cut through the jungle near the cage, but his head was concealed under a clump of leaves. All at once I caught a glimpse of some moving object at the edge of the path on the opposite side of the cage. It was a huge female gorilla, carrying a young one on her back. When I first saw her she was not more than thirty feet away. She was creeping along the edge of the bushes and watching the dog, who was busy with the bone. Her tread was so stealthy that I could not hear the rustle of a leaf. She advanced a few feet, crouched under the edge of the bushes, and cautiously peeped at the dog. She advanced again a little way, halted, crouched, and peeped again. It was evident that her purpose was to attack, and her approach was so wary as to leave no doubt of her dexterity in attacking a foe. Every movement was the embodiment of stealth. Her face wore a look of anxiety with a touch of ferocity. Her movements were quick but accurate, and her advance was not delayed by any indecision. The dog had not discovered her, and the smell of the bone and the noise he was making with it prevented his either smelling or hearing her. I could not warn him without alarming her. If he could have seen her before she made the attack, I should have left him to take his chances by flight or by battle. I should have been glad of an opportunity to witness such a combat and to study the actions of the belligerents, but I could not consent to see a friendly dog taken at such disadvantage. She was now rapidly covering the distance between them, and the dog had not yet discovered her. When she reached a point within about four yards of him I determined to break the silence. I cocked my rifle, and the click of the trigger caught her attention. I think this was the first thing that made her aware of my presence. She instantly stopped, turned her face and body towards the cage, and sat down on the ground in front of it. She gave me such a look that I almost felt ashamed of having interfered. She sat for fully one minute staring at me as if she had been transfixed. There was no trace of anger or of fear, but the look of surprise was on every feature. I could see her eyes move from my head to my feet. She scanned me as closely as if it had been her purpose to purchase me. At length she glanced at the dog, who was still eating the bone, then turned her head uneasily, as if to search for some way of escape. She rose, and retraced her steps with moderate haste; she did not run, but lost no time. She glanced back from time to time to see that she was not pursued. She uttered no sound of any kind.

From the time this ape came in view until she departed was about four minutes, and during that time I was afforded an opportunity of studying her in a way that no one else has ever been able to do. I watched every movement of her body, face and eyes. I could sit with perfect composure and study her without the fear of attack. With due respect for the temerity of men, I do not believe that any sane man could calmly sit and watch one of these huge beasts approach so near him without feeling a tremor of fear, unless he was protected as I was. Any man would either shoot or retreat, and he could not possibly study the subject with equanimity.

The temptation to shoot her was almost too great to resist, and the desire to capture her babe made it all the more so; but up to that time I had refrained from firing my gun anywhere within a radius of half a mile or so of my cage, and the natives had agreed to the same thing. My purpose in doing so was to avoid frightening the apes away from the locality. I had been told by the native hunters before this, that if I wounded one of them the others would leave the vicinity and not return perhaps for weeks. They say if you kill one the others do not appear to notice it so much as if it were wounded, although they seem to be aware of the fact and for the time flee, but will return again within a short time.

I could have shot this one with perfect ease and safety. As she approached, her head and breast were towards me; just before she discovered me her left side was in plain view, and when she sat down her breast was perfectly exposed, so that I could have shot her in the heart, the breast, or the head.

Her baby lay upon her back, with its arms embracing her neck and its feet caught under her arms. The cunning little imp saw me long before the mother did, but it gave her no warning of danger. It lay with its cheek resting on the back of her head. Its black face looked as smooth and soft as velvet. Its big brown eyes were looking straight at me, but it betrayed no sign of fear or even of concern. It really had a pleased expression, and was the nearest approach to a smile I have ever seen on the face of a gorilla. I believe that this is their method of carrying the young, and I have elsewhere assigned other reasons for this belief. In this case it is not a matter of belief, but one of knowledge, and everything that I have observed conspires to say that this is no exception to the rule.

During my sojourn of nearly four months in the jungle, where it was said the greatest number of gorillas could be found of any other place in the basin of that lake, I only saw a total of twenty-two, besides one other that I saw at another time in the forest while I was hunting. I only caught a glimpse of him, and should not even have done that had not the native guide discovered and pointed him out to me. I believe that no other white man has ever seen an equal number of these animals in a wild state, and it is certain that no other has ever seen them under as favourable conditions for study. I have compared notes with many white men on that part of the coast, but I have never found any reliable man who claims to have seen an equal number.

I know men there who have lived in that part for years, who frequently hunt in the forest for days at a time, and yet never saw a live gorilla. I met one man on my last voyage who has lived on the edge of the gorilla country forty-nine years, makes frequent journeys through the bush and along the watercourses in the interest of trade, and this man told me himself that in all that time he had never seen a wild gorilla. I would cite Mr. James A. Deemin as an expert woodsman, a cool, daring hunter, and I have enjoyed several hunts with him. He has travelled, traded, and hunted through the gorilla country for more than thirteen years, and has told me that with one exception he had never seen but one wild gorilla. This was a young one, and the exception alluded to was that he one time saw a school of them at a distance. On this occasion he was in a canoe and under the cover of the bushes along the side of a river until he came near them unobserved. Another man, whose name I will take the liberty of giving, is Mr. J. H. Drake, of Liverpool. Mr. Drake has never been suspected by those who know him of lacking courage in the hunt or being given to romance, and yet in many years on the coast he never saw but one school of these apes, and that was the same one that Mr. Deemin saw when they were travelling together. I could cite many others to show that it is a rare thing for the most expert woodsman ever to see one of these creatures, and many of the stories told by the casual traveller cannot be received with implicit faith. I do not mean to impeach the veracity of others, but fancy must have something to do with the case. While we cannot prove the negative by direct evidence, we must be permitted to doubt whether or not these apes are so frequently met in the jungle as they are alleged to be. I will give some reasons why I am a sceptic on this subject.

Almost every yarn told by the novice is quite the same in substance and much the same in detail as those related by others. It seems that most of them meet the same old gorilla, still beating his breast and screaming just as he did thirty years ago. The number of gun-barrels that he is accused of having chewed up would make an arsenal that would arm the volunteers. What becomes of all those that are attacked by this fierce monarch of the jungle? Not one of them ever gets killed, and not one of them ever kills a gorilla. Does he merely do this as a bluff and then recede from the attack? Or does he follow it up and seize his victim, tear him open and drink his blood as he is supposed to do? How does the victim escape? What becomes of the assailant? Who lives to tell the tale?

The gorilla has good ears, good eyes, and is a skilful bushman. One man walking through the jungle will make more noise than half a dozen gorillas. The gorilla can always see and hear a man before he is seen or heard by him. He is shy, and will not attack a man unless he is disturbed by him. He is always on the alert for danger, and rarely comes into the open parts of the bush except for food. He can conceal himself with more ease than a man can, and has every advantage in making his escape. I do not believe that he will ever approach a man if he can evade him. I quite believe that he will make a strong defence if surprised or attacked, but I do not believe it possible for any one to see a great number of gorillas in any length of time unless he goes to some one place and remains there as I have done. Even then he must sometimes wait for days without a trace of one. Silence and patience alone will enable him to see them; but when the gorilla sees him he at once retires as soon as he discovers the nature of the thing before him. He does not always flee in haste as many other animals do, but is more deliberate and cool. He will retreat in good order, and as a rule always starts in time if possible to escape without being observed. I trust that I may be pardoned for not being able to believe that every stranger who visits that country is attacked by a gorilla.

In addition to those I have seen in a wild state, I have seen about ten in captivity. Two of those were my own. They were good subjects for study, and I made the best use of them for the time I had them.

I accomplished one thing while in the jungle, for which I feel a just sense of pride, and that was making a gorilla take a portrait of himself. This will interest the amateur in the art of snapshots, and I shall relate it.

I selected a place in the forest where I found some tracks of the animal along the edge of a dense thicket of *batuna*. Under cover of the foliage I set up two pairs of stakes which were crossed at the tops, and to them was lashed a short pole forming something like a sawbuck. To this was fastened the camera, to which had been attached a trigger made of bamboo splits. One end of a string was fastened to the trigger, and the other end carried under a yoke to a distance of eight feet from the lens. At this point was attached a fresh plantain stalk and a nice bunch of the red fruit of the *batuna*. Upon this point the camera was focussed, the trigger was set, and it was left to await the gorilla. That afternoon I returned to find that something had taken the bait, broken the string, sprung the trigger and snapped the camera. I developed the plate, but could find no image of anything except the leaves in front of it. I repeated the experiment with the same results, but could not understand how anything could steal the bait and yet not be shown in the picture. The third time I did this I was gratified to find the image of a gorilla, and also to discover the cause why the others had not succeeded. The deep shadows of the forest make it difficult to take a photograph without giving it a time exposure, and when the sun is under a cloud or on the wrong side of an object it is quite impossible. The leaves that were shown in the first two plates were only those which were most exposed to the light, and all the lower part of the picture was without detail. In the third trial it could be seen that the sun was shining at the instant of exposure. A part of the body of the gorilla was in the light, but most of it was in the shadow of the leaves above it. The left side of the head and face were quite distinct, also the left shoulder and arm. The hand and bait could not have been distinguished except by their context. The right side of the head, arm, and most of the body were lost. The picture showed that he had taken the bait with his left hand, and that he was in a crouching posture at the moment. While the photograph was very poor as a work of art, it was full of interest as an experiment.

Although it did not result in getting a good picture, I do not regard the effort as a failure. It shows at least that such a thing is possible, and by careful efforts often repeated it could be made a means of obtaining some novel pictures. A little ingenuity would widen the scope of this device, and make it possible to photograph birds, elephants, and everything else in the forest. When I return to that place on a like journey, I shall carry the scheme into better effect.

CHAPTER XVII
OTHER APES

In the various records that constitute the history of these apes are found many novel and incoherent tales, but all of them appear to rest upon some basis of truth. In order to arrive at some more definite knowledge concerning them, we may review the data at our command. The first record in the annals of the world that alludes to these man-like apes, is that of Hanno, who made a voyage from Carthage to the west coast of Africa, nearly 500 years before the Christian era. He described an ape which was found in the locality about Sierra Leone. It is singular that the description which he gave of those apes should coincide so fully with those known of the present day, but to my mind it is quite certain that the ape of which he gives an account was neither a gorilla nor chimpanzee, nor is there anything to show that either of these ever occupied that part of the world, or that any similar type has done so. It is clear from the evidence that the ape described by him was not an anthropoid, but was the large, dog-faced monkey technically called *cynocephalus*. These animals are found all along the north coast of the Gulf of Guinea, but there is not a trace of any true ape along it north of Cameroon River, which empties into the sea about 4° north of the equator. Here begins the first trace of the chimpanzee. In passing along the windward coast, casual reports are current to the effect that gorillas and chimpanzees occupy the interior north of there; but when these reports are sifted down to solid facts, it always turns out to be a big baboon or monkey upon which the story rests. Its likeness to man as described by Hanno was doubtless the work of fancy, and the name *troglodytes* which he gave to it shows that he knew but little of its habits, or cared but little for the exactness of his statements.

The account given by Henry Battel, in 1590, contains a thread of truth woven into a web of fantasy. He must have heard the stories he relates, or seen the specimens along the coast north of the Congo, and there are certain facts which point to this conclusion. The name *pongo* which he gave to one of them belongs to the Fiot tongue, which is spoken by the native tribes around Loango. Those people apply the name to the gorilla, and is commonly understood to be synonymous with the name *njina*, used by the tribes north of there, and always applied to the gorilla. To me, however, it appears to coincide with the name *ntyii* as used by the Esyira people for another ape which is described in the chapter devoted to gorillas. It was from Loango that Dr. Falkenstein secured an ape under that name in 1876. It is singular that Baron Wurmb, in 1780, makes use of this same name *pongo* for an orang. I have not been able to learn where he acquired this name, but it appears to be a native Fiot name, and the history of their language is fairly well known for more than 400 years. The other name "Enjocko," given by

Battel to the other ape, is beyond a doubt a corruption of the native name *ntyigo* (ntcheego), and this name belongs north of the Congo from Mayumba to Gaboon. He may have inferred that these apes occupied Angola, but there is not a vestige of proof that any ape exists in that part of Africa. Even the native tribes of that part have no indigenous name for either one of these apes. Other parts of his account are erroneous, and while he may have believed that those apes "go in bodies to kill many natives that travel in the wood," and the natives may have told him such a thing, the apes do not practise such a habit. With all their sagacity they have no idea of the unity of action. If a band of them were attacked, they would no doubt act together in their defence, but it is not to be believed that they ever preconcert any plan of attack. Neither do these apes ever assault elephants. He is one animal they hold in mortal dread. I have incidentally mentioned elsewhere the conduct of my two kulus on board the ship when they saw a young elephant. Chico, the big ape that has also been mentioned, was often vicious and stubborn. Whenever he refused to obey his keeper or became violent, an elephant was brought in sight of his cage. On seeing it he became as docile as a lamb, and showed every sign of the most intense fear. Mr. Bailey himself told me of the dread both of his apes had for an elephant. Battel was also wrong in the mode he described of the mother carrying its young, and the apes using sticks or clubs.

The ape known as "Mafuka," which was exhibited in Dresden in 1875, was also brought from the Loango coast, and it is possible that this is the ape to which the native name *pongo* really belonged. This specimen in many respects conforms to the description of the *ntyii* given, but the idea suggested by certain writers that "Mafuka" was a cross between the gorilla and chimpanzee is not, to my mind, a tenable supposition. It would be difficult to believe that two apes of different species in a wild state would cross, but to believe that two that belonged to different genera would do so is even more illogical.

I may state here, however, again that some of the Esyira people advance such a theory concerning the *ntyii*, but the belief is not general, and those best skilled in woodcraft regard them as distinct species.

To quote, in pidjin English, the exact version of their relationship as it was given to me by my interpreter while in that country, may be of interest to the reader. I may remark, by way of explaining the nature of pidjin English, that it is a literal translation of the native mode of thought into English words. The statement was:

"*Ntyii* be one: *njina* be one: all two be one, one. *Nytii* 'e one mudder: *jnina* 'e one mudder: all two 'e one, one. *Nytii* 'e one fader: *njina* all same 'e one fader, 'e one. 'E all two one fader." By which the native means to say that the *nytii* has one mother and the *njina* has one mother, so that the two have two mothers, but both have one father, therefore they are half-brothers.

The other version given in denial of this statement was as follows:

"*Nytii* 'e one mudder: *njina* 'e one mudder. 'E one, one. *Nytii* 'e one fader: *njina* 'e one fader. 'E be one, one. All two 'e one, one. *Nytii* 'im mudder, *njina* 'im mudder. 'E brudder. *Nytii* 'im fader, *njina* 'im fader 'e brudder. All two 'e one, one."

The translation of this elegant speech is, that the *nytii* has a mother, and the *njina* has a mother which are not the same but sisters. The *nytii* has a father, and the *njina* has a father which are not the same, but are brothers, and therefore the two apes are only cousins, which in the native esteem is a remote degree of kinship.

The ape described by Lopez certainly belonged to the territory north of the Congo, which coast he explored, and gave his name to a cape about forty miles south of the equator, and it still bears the name Cape Lopez. At that time, however, it is probable that most of the low country now occupied by these apes was covered with water; that the lakes of that region were then all embraced in one great estuary, reaching from Fernan Vaz to Nazareth Bay, and extending eastward to the Foot hills below Lamberene. There is abundant evidence to show that such a state has once existed there, but it is not probable that these apes have ever changed their latitude.

The name "soko" appears to be a local name for the ordinary type of chimpanzee found throughout the whole range of their domain, and known in other parts by other names.

In Malimbu the name "kulu" appears to apply to the same species, while in the south-western part of their habitat that name, coupled with the verb "kamba," is confined strictly to the other type. Along the northern borders of the district to which that species belongs, but where he is very seldom found and little known to the natives, he is called Mkami tribe, "kanga ntyigo," to distinguish him from the common variety to which the latter name only is applied.

The etymology of the name *kanga* as applied to this ape is rather obscure. In common use it is a verb with the normal meaning to "parch" or "fry," and hence the secondary meaning to "prepare." Since this ape is said to be of a higher order of the race, the term is used to signify that he is "better prepared" than the other. That is to say, he is prepared to think and talk in a better manner.

Another history of this word appears to be more probable. The ape to which the name is applied lives between the Mkami country and the Congo, and the name is possibly a perversion of kongo, and implies the kind of *ntyigo* that lives towards the great river of that name. The etymology of African names is always difficult because there is no record of them, but many of them can be traced out with great precision, and some of them are unique.

The name M'Bouvé, as given by Du Chaillu, I have not been able to identify. In one part of the country I was told that the word meant the "chief" or head of a family. In another part it was said to mean something like an advocate or champion, and was only applied to one ape in a family group. The Rev. A. C. Goode, a zealous missionary who recently died near Batanga, was stationed for twelve years at Gaboon. During that time he travelled all through the Ogowe and Gaboon valleys. He was familiar with the languages of that part, and he explained the word in about the same way.

Whatever may be said concerning the veracity of Paul Du Chaillu, there is one thing that must be said to his credit. He gave to the world more knowledge of these apes than all other men put together had ever done before, and while he may have given a touch of colour to many incidents, and related some native yarns, he told a vast amount of valuable truth, and I can forgive him for anything which he may have misstated, except one. That is starting that story about gorillas chewing up gun-barrels. It has been a staple yarn in stock ever since, and the instant you ask a native any question about the habits of a gorilla he begins with this.

In view of the fact that I have made careful and methodic efforts to determine the exact boundary of the habitat and the real habits of these two apes, I feel at liberty to speak with an air of authority. I have acquired my knowledge on the subject by going to their own country and living in their own jungle, and I have thus obtained their secrets from first hands. With due respect to those who write books and speak freely upon subjects of which they know but little, I beg leave to suggest that if the authors had gone into the jungle and lived among those animals instead of consulting others who know less than themselves about it, many of them would have written in a very different strain. I do not mean this as a rebuke to any one, but seeing the same old stories repeated year after year, and knowing that there is no truth in them, I feel it incumbent as a duty to challenge them.

I believe that in the future it will be shown that there are two types of gorilla as distinct from each other as the two chimpanzees now known. This second variety of gorilla will be found between the third and fifth parallels south and east of the delta district, but west of the Congo. I believe it was represented in the ape "Mafuka." My researches among the apes have been confined chiefly to the two kinds heretofore described, but I have seen and studied in a superficial way the orang and the gibbon. I am not prepared as yet to discuss the habits of those two apes, but as they form a part of the group of anthropoids we cannot dismiss them without honourable mention.

The orang-outan, as he is called in his own country, is known to zoology by the first of these terms alone. He is a native of Borneo and Sumatra, and opinions differ as to whether there are two species or only one.

The general plan of the skeleton of the orang is very much the same as in the other apes. The chief points of difference are that it has one bone more in the wrist and one joint less in the spinal column than is found in man. He has thirteen pairs of ribs, which appear to be more constant in their number than in man. His arms are longer and his legs shorter in proportion to his body than the other two apes. The type of the skull is peculiar, and combines to a certain extent more human-like form in one part with a more beast-like form in another. The usual height of an adult male is about fifty-one inches.

I have never had an opportunity of studying this ape in a wild state, and have only had access to four of them in captivity, all of which were young and most of them inferior specimens. He is the most obtuse or stupid of the four great apes. And were it not for his skeleton alone he would be assigned a place below the gibbon, for in point of speech and mental calibre he is far inferior. The best authorities perhaps upon the habits of this ape in a wild state are Messrs. W. T. Horniday and R. A. Wallace.

The first and last in order of the anthropoid apes is the gibbon; he is much smaller in size, greater in variety, and more active than any other of the group. His habitat is in the south-east of Asia; its outline is vaguely defined, but it includes the Malayan Peninsula and many of the contiguous islands east and south of it.

The skeleton of the gibbon is the most delicate and graceful in build of all the apes, and in this respect is as far superior to man as man is to the gorilla, except for the long arms and digits. He is the only one of the four that can walk in an erect position, but in doing this the gibbon is awkward, and often uses his arms to balance himself, sometimes by touching his hands to the ground, or at other times raising them above his head or extending them on either side. The length of them is such that he can touch the fingers to the ground while the body is nearly if not quite erect. In the spinal column he has two and sometimes three sections more than man. His digits are very much longer, but his legs are nearly the same length in proportion to his body as those of man. He has fourteen pairs of ribs.

The gibbon is the most active, if not the most intelligent, of all apes. He is more arboreal in habit than any other. Many wonderful stories are told of his agility in climbing and leaping from limb to limb. One authentic report credits one of these apes with leaping a distance of forty-two feet from the limb of one tree to that of another. Perhaps a better term is to call it swinging rather than leaping, as these flights are performed by the arms. Another account is, that one swinging by one hand propelled himself a horizontal distance of eighteen feet through the air, seizing a bird in flight, and alighting safely upon another limb with his prey in hand.

There are several of this ape known, the largest of which is about three feet high, but the usual height is not more than thirty inches. The voice of one species is remarkable for its strength, scope and quality above all other apes. Most of the members of this genus are endowed with better vocal qualities than other animals. This ends the list of the man-like apes, and next in order after them come the monkeys, but we will deal with that subject more at length at some future time.

The descent, as we have elsewhere observed, from the highest ape to the lowest monkey presents one unbroken scale of imbricating planes; and we have seen in what degree man is related to the higher ape. From whence we may discern in what degree his physical nature is the same as that of all the order to which he belongs. No matter in what respect he may differ in his mental and moral nature, his likeness to them should at least restrain his pride, evoke his sympathy, and share the bounty of his benevolence. Let man realise to its full extent that he is one in nature with the rest, and they will receive the benign influence of his dignity without impairing it, while he will elevate himself by having given it.

CHAPTER XVIII
THE TREATMENT OF APES IN CAPTIVITY

In conclusion, I deem it in order to offer a few remarks with regard to the causes of death among these apes, and to the proper treatment of the animals in captivity. We know so little and assume so much concerning them that we often violate the very laws under which they live.

We have already noticed the fact that the gorilla is confined by nature to a low, humid region, reeking with miasma and the effluvia of decaying vegetation. The atmosphere in which he thrives is one in which human life can hardly exist. We know in part why man cannot live in such an atmosphere and under such conditions, but we cannot say with certainty why the ape does do so. It would seem that the very element that is fatal to the life of man gives strength and vitality to the gorilla.

We know that all forms of animal life are not affected in the same way by the same things, and while it may be said in round numbers that whatever is good for man is good for apes also, it is not a fact.

The human race is the most widely distributed of any genus of mammals and, as a race, can undergo the greatest extremes of change in climate, food and other conditions of any other animal. His migratory habits, both inherent and acquired, have fitted him for a life of vicissitudes, and such a life inures him as an individual to all extremes. On the other hand, the gorilla, as a genus, is confined to a small habitat, which is uniform in climate, products and topography; and having been so long restricted to these conditions he is unfitted for like changes, and when such are forced upon him the result must always be to his injury.

In certain parts of the American tropics there is found a rich, grey moss growing in great profusion in certain localities and on certain kinds of trees. It is not confined to any certain level, but thrives best on the lowest elevations. Under favourable conditions it will grow at altitudes far above the surrounding swamps. The character and quantity, however, are measured by the altitude at which it grows. It is an aerial plant, and may be detached from the boughs of one tree and transplanted upon those of another. It may be taken with safety for a great distance so long as an atmosphere is supplied to it that is suited to its nature; but when removed from its normal conditions and placed in a purer air it begins to languish and soon dies. If it be returned in time, however, to its former place or one of like character it will revive and continue to grow.

What element this plant extracts from the impure air is a matter of doubt; but it cannot be carbonic acid gas which is the chief food of plants, nor it cannot be any form of nitrogen; and it is well known that the plant cannot long survive in a pure atmosphere. Whatever the ingredient extracted may be, it is certain that it is one that is deadly to human life, and one which other plants refuse. Moisture and heat alone cannot account for it.

We have another striking instance in the eucalyptus, which lives upon the poison of the air around it. There are many other cases in vegetable life, and while the animal is a higher organism than the plant, there are certain laws of life that obtain in both kingdoms which are the same in principle.

Between the case of the gorilla and that of the plant there is some analogy. It may not be the same element that sustains them both, but it is possible that the very microbes which germinate disease and prove fatal to man sustain the life of the ape in the prime of health. The poison which destroys life in man preserves it in the ape.

The chimpanzee is distributed over a much greater range, and is capable of undergoing a much greater degree of change in food and temperature. The history of these apes in captivity shows that the chimpanzee lives much longer in that state and requires much less care. From my own observation I assert that all of these apes can undergo a greater range of temperature than they can of humidity. This appears to be one of the essential things to the life of a gorilla, and one fatal mistake made in treating him is furnishing him with a dry, warm atmosphere, and depriving him of the poison contained in the malarious air in which he spends his entire life. Both of these apes need humidity. The chimpanzee will live longer than a gorilla in a dry air, but neither of them can long survive it, and it would appear that a salt atmosphere is best for the gorilla.

I believe that one of these apes could be kept in good condition for any length of time if he were supplied with a normal humidity in an atmosphere laden with miasma and allowed to vary in temperature. A constant degree of heat is not good for any animal, there is nowhere in all the earth that nature sustains a uniform degree of it. We need not go to either extreme, but a change is requisite to bring into play all the organs of the body.

The theory of their treatment which I would advance is to build them a house entirely apart from that of any other animal. It should be 18 or 20 feet wide by 35 or 40 long, and at least 15 feet high. It should have no floor except earth, and that should be of sandy loam or vegetable mould. In one end of this building there should be a pool of water 12 or 15 feet in diameter, and embedded in the mould under the water should be a steam coil to regulate the temperature as might be desired. In this pool should be grown a dense crop of water plants such as are found in the marshes of the country in which the gorilla lives. This pool should not be cleaned out or the water changed, but the plants should be allowed to grow and decay in a natural way. Neither the pool nor the house should be kept at a uniform heat, but allowed to vary from 60 to 90 degrees.

In addition to the things mentioned, the place should be provided with the means of giving it a spray of tepid water, which should be turned on once or twice a day, and allowed to continue for at least an hour at a time. The water for this purpose should be taken from the pool, but should never be warmer than the usual temperature of tropical rain. The animal should not be required to take a bath in this way, but should be left to his own choice about it.

The house should be separated by a thin partition that could be removed at will, and the other end of the building from the pool should be occupied by a strong tree, either dead or alive, to afford the inmates proper exercise. The rule that visitors or strangers should not annoy or tease them should be enforced without respect to person, time, or rank. No visitor should be allowed on any terms to give them any kind of food. The reasons for these precautions are obvious to any one familiar with the keeping of animals, but in the case of a gorilla their observance cannot be waived with impunity.

The south side of the house should be of glass, and at least half of the top should be of the same. These parts should be provided with heavy canvas curtains, to be drawn over them so as to adjust or regulate the sunlight. In summer-time the building should be kept quite open so as to admit air and rain. The ape does not need to be pampered: on the contrary, he should be permitted to rough it. Half of the gorillas that have ever been in captivity have died from over-nursing. By nature they are strong and robust if the proper conditions are supplied, but when these are changed he becomes a frail and tender creature. They should not be restricted to a vegetable diet nor limited to a few articles of food of any kind, but should be allowed to select such things as they prefer to eat. I have grave doubts as to the wisdom of limiting the quantity. One mistake is often committed in the treatment of animals, and that is to continue the same diet at all times and limit that to one or two items. It may be observed that the higher the form of organism is the more diverse the taste becomes, and while very hardy animals or those of low forms may be restricted to one staple kind of food, the higher forms demand a change.

One thing above all others that I would inhibit is the use of straw of any kind in their cage for beds or any other purpose. If it be desired to furnish them with such a comfort, nothing should ever be used but dead leaves if they can be supplied. In their absence a canvas hammock or wire matting should be used. There are certain kinds of dust given off by the dry straw of all cereal plants. This is deleterious to the health of man, but vastly more so to these apes. It is taken into the lungs, and through them act upon other parts of the body by suppressing the circulation and respiration. No matter how clean the straw may be, the effect will be the same in the end. Hay is better than straw, but even this should not be used.

Another thing which is necessary is to entertain or amuse them in some way, otherwise they become despondent and gloomy. It is believed by those who are familiar with these apes that loneliness or solitude is a fruitful cause of death. This is especially so with the gorilla. I have a photograph of one that was kept by a trader on the coast of Africa for nearly three years. She was devoted to him, and was never content when not in his company. His business required him to make a journey of a few days to the interior. He left the gorilla at his place on the coast where she had lived up to this time. The day after he departed she became morose and fretful, and within a few days died without any apparent cause except pining. This was observed by natives and by white traders, and her death has always been ascribed to the cause assigned. She was well known to all the traders on that part of the coast, and has been regarded as one of the best specimens known. She is the only one that I have ever known to become devoted to a human being.

Another important fact that is little known but very singular is, that tobacco smoke is absolutely fatal to a gorilla. Every native hunter that I met in Africa testifies that this simple thing will kill any gorilla in the forest if he is subjected to the fumes for a short time. I have reason to believe that it is true. It may not prove fatal in every instance, but it will in many. The chimpanzee is not so much affected by it, although he dislikes it, but the gorilla detests it and shows at all times his strong aversion to it. I have no doubt that this is one of the reasons that these apes always die on board the ships by which they are brought from Africa.

Both of these apes are possessed, in a degree, of savage and resentful instincts. But these are much stronger in the gorilla than in the chimpanzee. He therefore requires firm and consistent treatment. This can be used without being severe or cruel, but the intellect of the gorilla must not be underrated. He studies the motives and intentions of man with a keen perception, and is seldom mistaken in his interpretation of them. He often manifests a violent dislike for certain persons, and when such is discovered to be the case the object of his dislike should not be permitted in his presence, for the result is to enrage the ape and excite his nervous nature. When they become sullen or obstinate they should not be coaxed or indulged, nor yet used with harshness. They should either be left alone for the time or diverted by a change of treatment.

At this point I submit the foregoing to the world as the sum of my labours in this special field of research up to this time. I regret that I have been compelled to deny much that has been said, but I make no apology for having done so. In this work I have sought to place these apes before the reader as I have seen them in their native forest. I have not clothed them in fine raiment or invested them in glamour, but I trust that this contribution may be found worthy of the respect of all men who love Nature and respect fidelity.

I have the vanity to believe that the methods of study which I have employed will be made the means of farther research by more able students than the writer.

www.ingramcontent.com/pod-product-compliance
Lightning Source LLC
Chambersburg PA
CBHW021933190326
41519CB00009B/1004